Essential Mathematics
With Applications

Essential Mathematics

With Applications

THIRD EDITION

Vernon C. Barker
Palomar College, California

Richard N. Aufmann
Palomar College, California

HOUGHTON MIFFLIN COMPANY   Boston
Dallas   Geneva, Illinois
Palo Alto   Princeton, New Jersey

Chapter Opener Art designed and rendered by Daniel P. Derdula.

All Interest Grabber Art designed and rendered by Daniel P. Derdula (with airbrushing by Linda Phinney on illustration for Chapter 6).

All Interior Math Figures rendered by NETWORKGRAPHICS (135 Fell Court, Hauppauge, New York 11788).

Interior design by George McLean.

Cover photograph by Slide Graphics of New England, Inc.

Copyright © 1991 by Houghton Mifflin Company. All rights reserved.

No part of the format or content of this work may be reproduced or transmitted in any form or by any means, electronic or mechanical, including photocopying and recording, or by any information storage or retrieval system without the prior written permission of Houghton Mifflin Company unless such copying is expressly permitted by federal copyright law. Address inquiries to College Permissions, Houghton Mifflin Company, One Beacon Street, Boston, MA 02108.

Printed in the U.S.A.

ISBN Numbers:
Text: 0-395-54439-4
Answer Booklet: 0-395-57083-2

DEFGHIJ-VH-98765432

Contents

Preface *xi*

1 Whole Numbers *1*

SECTION 1.1 **Introduction to Whole Numbers** *3*

Objective A To identify the order relation between two numbers *3*
Objective B To write whole numbers in words and in standard form *4*
Objective C To write whole numbers in expanded form *4*
Objective D To round a whole number to a given place value *5*

SECTION 1.2 **Addition of Whole Numbers** *9*

Objective A To add whole numbers without carrying *9*
Objective B To add whole numbers with carrying *10*
Objective C To solve application problems *12*

SECTION 1.3 **Subtraction of Whole Numbers** *19*

Objective A To subtract whole numbers without borrowing *19*
Objective B To subtract whole numbers with borrowing *20*
Objective C To solve application problems *22*

SECTION 1.4 **Multiplication of Whole Numbers** *27*

Objective A To multiply a number by a single digit *27*
Objective B To multiply larger whole numbers *28*
Objective C To solve application problems *30*

SECTION 1.5 **Division of Whole Numbers** *35*

Objective A To divide by a single digit with no remainder in the quotient *35*
Objective B To divide by a single digit with a remainder in the quotient *37*
Objective C To divide by larger whole numbers *39*
Objective D To solve application problems *41*

SECTION 1.6 **Exponential Notation and the Order of Operations Agreement** *47*

Objective A To simplify expressions containing exponents *47*
Objective B To use the Order of Operations Agreement to simplify expressions *48*

SECTION 1.7 **Prime Numbers and Factoring** *51*

    Objective A   To factor numbers   *51*
    Objective B   To find the prime factorization of a number   *52*

Calculators and Computers   *55*
Chapter Summary   *55*
Chapter Review   *57*
Chapter Test   *59*

2 Fractions *61*

SECTION 2.1 **The Least Common Multiple and Greatest Common Factor** *63*

    Objective A   To find the least common multiple (LCM)   *63*
    Objective B   To find the greatest common factor (GCF)   *64*

SECTION 2.2 **Introduction to Fractions** *67*

    Objective A   To write a fraction that represents part of a whole   *67*
    Objective B   To write an improper fraction as a mixed number or a whole number, and a mixed number as an improper fraction   *68*

SECTION 2.3 **Writing Equivalent Fractions** *71*

    Objective A   To find equivalent fractions by raising to higher terms   *71*
    Objective B   To write a fraction in simplest form   *72*

SECTION 2.4 **Addition of Fractions and Mixed Numbers** *75*

    Objective A   To add fractions with the same denominator   *75*
    Objective B   To add fractions with unlike denominators   *75*
    Objective C   To add whole numbers, mixed numbers, and fractions   *76*
    Objective D   To solve application problems   *78*

SECTION 2.5 **Subtraction of Fractions and Mixed Numbers** *83*

    Objective A   To subtract fractions with the same denominator   *83*
    Objective B   To subtract fractions with unlike denominators   *83*
    Objective C   To subtract whole numbers, mixed numbers, and fractions   *84*
    Objective D   To solve application problems   *86*

SECTION 2.6 **Multiplication of Fractions and Mixed Numbers** *91*

    Objective A   To multiply fractions   *91*
    Objective B   To multiply whole numbers, mixed numbers, and fractions   *92*
    Objective C   To solve application problems   *94*

SECTION 2.7 **Division of Fractions and Mixed Numbers** *99*

    Objective A   To divide fractions   *99*

Objective B   To divide whole numbers, mixed numbers, and fractions   99
Objective C   To solve application problems   101

SECTION 2.8   **Order, Exponents, and the Order of Operations Agreement**   107

Objective A   To identify the order relation between two fractions   107
Objective B   To simplify expressions containing exponents   107
Objective C   To use the Order of Operations Agreement to simplify expressions   108

Calculators and Computers   *111*
Chapter Summary   *111*
Chapter Review   *113*
Chapter Test   *115*
Cumulative Review   *117*

3 Decimals   119

SECTION 3.1   **Introduction to Decimals**   121

Objective A   To write decimals in standard form and in words   121
Objective B   To round a decimal to a given place value   122

SECTION 3.2   **Addition of Decimals**   125

Objective A   To add decimals   125
Objective B   To solve application problems   126

SECTION 3.3   **Subtraction of Decimals**   129

Objective A   To subtract decimals   129
Objective B   To solve application problems   130

SECTION 3.4   **Multiplication of Decimals**   133

Objective A   To multiply decimals   133
Objective B   To solve application problems   136

SECTION 3.5   **Division of Decimals**   141

Objective A   To divide decimals   141
Objective B   To solve application problems   144

SECTION 3.6   **Comparing and Converting Fractions and Decimals**   149

Objective A   To convert fractions to decimals   149
Objective B   To convert decimals to fractions   149
Objective C   To identify the order relation between two decimals or between a decimal and a fraction   150

Calculators and Computers   *153*
Chapter Summary   *153*
Chapter Review   *155*

Chapter Test *157*
Cumulative Review *159*

4 Ratio and Proportion *161*

SECTION 4.1 **Ratio** *163*

    Objective A   To write the ratio of two quantities in simplest form *163*
    Objective B   To solve application problems *164*

SECTION 4.2 **Rates** *167*

    Objective A   To write rates *167*
    Objective B   To write unit rates *167*
    Objective C   To solve application problems *168*

SECTION 4.3 **Proportions** *171*

    Objective A   To determine if a proportion is true *171*
    Objective B   To solve proportions *172*
    Objective C   To solve application problems *174*

Calculators and Computers *179*
Chapter Summary *180*
Chapter Review *181*
Chapter Test *183*
Cumulative Review *185*

5 Percents *187*

SECTION 5.1 **Introduction to Percents** *189*

    Objective A   To write a percent as a fraction or decimal *189*
    Objective B   To write a fraction or a decimal as a percent *190*

SECTION 5.2 **Percent Equations: Part I** *193*

    Objective A   To find the amount when the percent and base are given *193*
    Objective B   To solve application problems *194*

SECTION 5.3 **Percent Equations: Part II** *197*

    Objective A   To find the percent when the base and amount are given *197*
    Objective B   To solve application problems *198*

SECTION 5.4 **Percent Equations: Part III** *201*

    Objective A   To find the base when the percent and amount are given *201*

Objective B   To solve application problems   201

SECTION 5.5   **Percent Problems: Proportion Method**   205

Objective A   To solve percent problems using proportions   205
Objective B   To solve application problems   206

Calculators and Computers   209
Chapter Summary   210
Chapter Review   211
Chapter Test   213
Cumulative Review   215

6 Applications for Business and Consumers: A Calculator Approach   217

SECTION 6.1   **Applications to Purchasing**   219

Objective A   To find unit cost   219
Objective B   To find the most economical purchase   219
Objective C   To find total cost   220

SECTION 6.2   **Percent Increase and Percent Decrease**   223

Objective A   To find percent increase   223
Objective B   To apply percent increase to business—markup   224
Objective C   To find percent decrease   226
Objective D   To apply percent decrease to business—discount   227

SECTION 6.3   **Interest**   233

Objective A   To calculate simple interest   233
Objective B   To calculate compound interest   234

SECTION 6.4   **Real Estate Expenses**   237

Objective A   To calculate the initial expenses of buying a home   237
Objective B   To calculate ongoing expenses of owning a home   238

SECTION 6.5   **Car Expenses**   243

Objective A   To calculate the initial expenses of buying a car   243
Objective B   To calculate ongoing expenses of owning a car   244

SECTION 6.6   **Wages**   247

Objective A   To calculate commissions, total hourly wages, and salaries   247

SECTION 6.7   **Bank Statements**   251

Objective A   To calculate checkbook balances   251
Objective B   To balance a checkbook   252

Calculators and Computers 261
Chapter Summary 261
Chapter Review 263
Chapter Test 265
Cumulative Review 267

APPENDIX A1 Addition Table *A2*
Multiplication Table *A2*
Table of Square Roots *A3*
Compound Interest Table *A4*
Monthly Payment Table *A6*

Solutions to Student Examples *A7*

Answers to Odd-Numbered Exercises *A29*

Index *A41*

Preface

The third edition of Essential Mathematics With Applications provides mathematically sound and comprehensive coverage of the topics considered essential in a basic mathematics course. Our strategy in preparing this revision has been to build on the successful features of the second edition, features designed to enhance the student's mastery of math skills.

Features

The Interactive Approach

Instructors have long recognized the need for a text that requires the student to use a skill as it is being taught. Essential Mathematics With Applications uses an interactive technique that meets this need. Each section is divided into objectives, and every objective contains one or more sets of matched-pair examples. The first example in each set is worked out; the second example is not. By solving this second problem, the student interacts with the text. The complete worked-out solutions to these examples are provided in an appendix at the end of the book, so the student can obtain immediate feedback on and reinforcement of the skill being learned.

Emphasis on Problem-Solving Strategies

Essential Mathematics With Applications features a carefully developed approach to problem solving that emphasizes developing strategies to solve problems. For each type of word problem contained in the text, the student is prompted to use a "strategy step" before performing the actual manipulation of numbers. By developing problem-solving strategies, the student will know better how to analyze and solve those word problems encountered in a variety of courses.

Applications

The traditional approach to teaching or reviewing mathematics, which places major emphasis on the manipulation of numbers, is lacking in that it fails to teach students the practical value of mathematics. By contrast, Essential Mathematics With Applications emphasizes applications. Wherever appropriate, the last objective in each section presents applications that require the student to use the skills covered in that section to solve practical problems. Also, all of Chapter 6, "Applications for Business and Consumers: A Calculator Approach," and portions of several other chapters are devoted entirely to applications. This carefully-integrated applied approach generates awareness on the student's part of the value of mathematics as a real-life tool.

Keyed Objectives

Each chapter begins with a list of the learning objectives included within that chapter. Each of the objectives is then restated in the chapter to remind the student of the current topic of discussion.

Exercises

There are more than 3000 exercises in the text, grouped in the following categories:

- **End-of-section exercise sets,** which are keyed to the corresponding learning objectives, provide ample practice and review of each skill.
- **Chapter review exercises,** which appear at the end of each chapter, help the student integrate all of the skills presented in the chapter.
- **Chapter tests,** which appear at the end of each chapter, are typical one-hour exams that the student can use to prepare for an in-class test.
- **Cumulative review exercises,** which appear at the end of each chapter (beginning with Chapter 2), help the student retain math skills learned in earlier chapters.
- **Calculator exercises,** which appear in the end-of-section exercises, are included throughout the text. These exercises provide the student with the opportunity to practice using a hand-held calculator and are identified by a special color box printed over the exercise number for each calculator exercise.

Calculator and Computer Enrichment Topics

Each chapter also contains an optional calculator or computer enrichment topic. Calculator topics provide the student with valuable key-stroking instructions and practice in using a hand-held calculator. Computer topics correspond directly to the programs found on the Math ACE (Additional Computer Exercises) Disk. These topics range from adding and subtracting fractions to calculating monthly mortgage payments.

New To This Edition

Topical Coverage

The topic of Estimation has been added to Chapters 1 and 3, along with an abundance of Estimation exercises.

In this edition Chapter 6, "Statistics," includes new material on reading pictographs.

In keeping with our commitment to applications, over one hundred word problems have been rewritten to reflect contemporary situations.

Chapter Review and Chapter Test

There is now a Chapter Review and a Chapter Test at the end of each chapter. The Chapter Review is organized by section with the use of section heads; the Chapter Test is organized by objectives. The objective references for the Chapter Test do not appear on the test page, so students are not prompted in any way. Instead, the objective references are given in the Answer Section at the back of the book. Thus the student can refer back to the Answer Section to find out which objective to restudy if necessary.

Supplements For The Student

Two computerized study aids, the Computer Tutor™ and the Math ACE (Additional Computer Exercises) Disk have been carefully designed for the student.

The same objectives that organize the text organize each computerized ancillary.

The COMPUTER TUTOR™

The Computer Tutor™ is an interactive instructional microcomputer program for student use. Each learning objective in the text is supported by a lesson on the Computer Tutor™. As a reminder of this, a small computer icon appears to the right of each objective title in the text. Lessons on the tutor provide additional instruction and practice and can be used in several ways: (1) to cover material the student missed because of absence from class; (2) to repeat instruction on the skill or concept that the student has not yet mastered; or (3) to review material in preparation for examinations. This tutorial program is available for the IBM PC and compatible computers.

Math ACE (Additional Computer Exercises) Disk

The Math ACE Disk contains a number of computational and drill-and-practice programs that correspond to selected Calculator and Computer Enrichment Topics in the text. These programs are available for the IBM PC and compatible computers.

Supplements for the Instructor

Instructor's Answer Booklet

The Instructor's Answer Booklet contains the answers to all end-of-section exercise sets, Chapter Reviews, Chapter Tests, and Cumulative Reviews.

Acknowledgements

The authors would like to thank the people who have reviewed this manuscript and provided many valuable suggestions:

Geoffrey Akst
Borough of Manhattan Community College, NY

Betty Jo Baker
Lansing Community College, MI

Judith Brower
North Idaho College, ID

Mary Cabral
Middlesex Community College, MA

Carmy Carranga
Indiana University of Pennsylvania, PA

Patricia Confort
Roger Williams College, RI

Michael Contino
California State University, CA

Bob C. Denton
Orange Coast College, CA

Joel Greenstein
New York Technical College, NY

Bonnie-Lou Guertin
Mt. Wachusett Community College, MA

Lynn Hartsell
Dona Ana Branch Community College, NM

Ida E. Hendricks
Shepherd College, WV

Diana L. Hestwood
Minneapolis Community College, NM

Pamela Hunt
Paris Junior College, TX

Arlene Jesky
Rose State College, OK

Sue Korsak
New Mexico State University, NM

Randy Leifson
Pierce College, WA

Virginia M. Licata
Camden County College, NJ

Judy Liles
North Harris County College–South Campus, TX

David Longshore
Victor Valley College, CA

Margaret Luciano
Broome Community College, NY

Carl C. Maneri
Wright State University, OH

Rudy Maglio
DePaul University, IL

Patricia McCann
Franklin University, OH

Dale Miller
Highland Park Community College, MI

Judy Miller
Delta College, MI

Scott L. Mortensen
Dixie College, UT

Michelle Mosman
Des Moines Area Community College, IA

Wendell Neal
Houston Community College, TX

Doris Nice
University of Wisconsin–Parkside, WI

Kent Pearce
Texas Tech University, TX

Sue Porter
Davenport College, MI

Jack Rotman
Lansing Community College, MI

Karen Schwitters
Seminole Community College, FL

Dorothy Smith
Del Mar College, TX

William N. Thomas, Jr.
University of Toledo, OH

Dana Mignogna Thompson
Mount Aloysius Junior College, PA

Paul Treuer
University of Minnesota–Duluth, MN

Thomas Wentland
Columbus College, GA

William T. Wheeler
Abraham Baldwin College, GA

Harvey Wilensky
San Diego Miramar College, CA

Wayne Wolfe
Orange Coast College, CA

Kenneth Word
Central Texas College, TX

Justane Valudez-Ortiz
San Jose College, CA

To the Student

Many students feel that they will never understand math while others appear to do very well with little effort. Often times what makes the difference is that successful students take an active role in the learning process.

Learning mathematics requires your *active* participation. Although doing homework is one way you can actively participate, it is not the only way. First, you must attend class regularly and become an active participant. Secondly, you must become actively involved with the textbook.

Essential Mathematics With Applications was written and designed with you in mind as a participant. Here are some suggestions on how to use the features of this textbook.

There are 6 chapters in this text. Each chapter is divided into sections and each section is subdivided into learning objectives. Each learning objective is labeled with a letter from A–E.

First, read each objective statement carefully so you will understand the learning goal that is being presented. Next, read the objective material carefully, being sure to note each bold word. These words indicate important concepts that you should familiarize yourself with. Study each in-text example carefully, noting the techniques and strategies used to solve the example.

You will then come to the key learning feature of this text, the *boxed examples*. These examples have been designed to aid you in a very specific way. Notice that in each example box, the example on the left is completely worked out and the example on the right is not. The reason for this is that *you* are expected to work the right-hand example (in the space provided) in order to immediately test your understanding of the material you have just studied.

You should study the worked-out example carefully by working through each step presented. This allows you to focus on each step and reinforces the technique for solving that type of problem. You can then use the worked-out example as a model for solving similar problems.

Next, try to solve the right-hand example using the problem-solving techniques that you have just studied. When you have completed your solution, check your work by turning to the page in the appendix where the complete solution can be found. The page number on which the solution appears is printed at the bottom of the example box in the right-hand corner. By checking your solution, you will know immediately whether or not you fully understand the skill just studied.

When you have completed studying an objective, do all of the exercises in the exercise set that correspond to that objective. The exercises will be labeled with the same letter as the objective. Math is a subject that needs to be learned in small sections and practiced continually in order to be mastered. Doing all of the exercises in each exercise set will help you master the problem-solving techniques necessary for success.

Once you have completed the exercises to an objective, you should check your answers to the odd-numbered exercises with those found in the back of the book.

After completing a chapter, read the Chapter Summary. This summary highlights the important topics covered in the chapter. Following the Chapter Summary are Chapter Review Exercises, a Chapter Test, and a Cumulative Review (beginning with Chapter 2). Doing the review exercises is an important way of testing your understanding of the chapter. The answer to each review exercise is given in an appendix at back of the book. Each answer is followed by a reference that tells which objective that exercise was taken from. For example, (4.2B) means Section 4.2, Objective B. After checking your answers, restudy any objective that you missed. It may be very helpful to retry some of the exercises for that objective to reinforce your problem solving techniques.

The Chapter Test should be used to prepare for an exam. We suggest that you try the Chapter Test a few days before your actual exam. Take the test in a quiet place and try to complete the test in the same amount of time you will be allowed for your exam. When taking the Chapter Test, practice the strategies of successful test takers: 1) scan the entire test to get a feel for the questions; 2) read the directions carefully; 3) work the problems that are easiest for you first; and perhaps most importantly, 4) try to stay calm.

When you have completed the Chapter Test, check your answers. If you missed a question, review the material in that objective and rework some of the exercises from that objective. This will strengthen your ability to perform the skills in that objective.

The Cumulative Review allows you to refresh the skills you learned in previous chapters. This is very important in mathematics. By consistently reviewing previous material, you will retain the skills already learned as you build new ones.

Remember, to be successful: attend class regularly; read the textbook carefully; actively participate in class; work with your textbook using the boxed examples for immediate feedback and reinforcement of each skill; do all the homework assignments; review constantly; and work carefully.

1 Whole Numbers

OBJECTIVES

- To identify the order relation between two numbers
- To write whole numbers in words and in standard form
- To write whole numbers in expanded form
- To round a whole number to a given place value
- To add whole numbers without carrying
- To add whole numbers with carrying
- To solve application problems
- To subtract whole numbers without borrowing
- To subtract whole numbers with borrowing
- To solve application problems
- To multiply a number by a single digit
- To multiply larger whole numbers
- To solve application problems
- To divide by a single digit with no remainder in the quotient
- To divide by a single digit with a remainder in the quotient
- To divide by larger whole numbers
- To solve application problems
- To simplify expressions containing exponents
- To use the Order of Operations Agreement to simplify expressions
- To factor numbers
- To find the prime factorization of a number

Family Tree for Numbers

Our number system is called the Hindu-Arabic system because it has its ancestry in India and was refined by the Arabs. But despite the influence of these cultures on our system, there is some evidence that our system may have originated in China around 1400 B.C. That is 34 centuries ago.

The family tree below illustrates the most popular belief of the history of our number system. In the 16th century, symbols for our numbers started to become standardized with the advent, by Gutenberg, of the printing press.

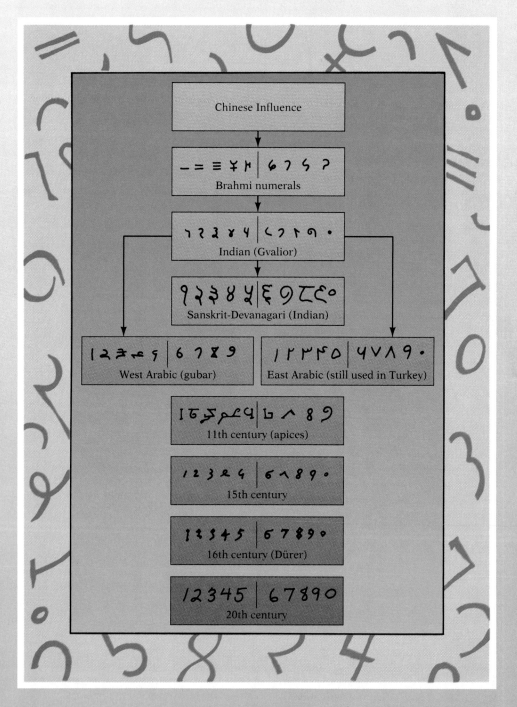

SECTION 1.1 Introduction to Whole Numbers

Objective A To identify the order relation between two numbers

The **whole numbers** are 0, 1, 2, 3, 4, 5, 6, 7, 8, 9, 10, 11, 12, 13, 14, . . .

The three dots mean that the list continues on and on and that there is no largest whole number.

Just as distances are associated with the markings on the edge of a ruler, the whole numbers can be associated with points on a line. This line is called the **number line.**

The Number Line

The **graph** of a whole number is shown by placing a heavy dot on the number line directly above the number.

The graph of 7 on the number line.

The number line can be used to show the order of whole numbers. A number that appears to the left of a given number is **less than** the given number. The symbol for "is less than" is $<$. A number that appears to the right of a given number is **greater than** the given number. The symbol for "is greater than" is $>$.

Four is less than seven.
$4 < 7$

Twelve is greater than seven.
$12 > 7$

Example 1 Graph 11 on the number line.

Solution

Example 2 Graph 9 on the number line.

Your solution

Example 3 Place the correct symbol, $<$ or $>$, between the two numbers.
   a. 39 24
   b.   0 51

Solution a. $39 > 24$
         b. $0 < 51$

Example 4 Place the correct symbol, $<$ or $>$, between the two numbers.
   a. 45 29
   b. 27   0

Your solution a.
              b.

Solutions on p. A7

Objective B To write whole numbers in words and in standard form

When a whole number is written using the digits 0, 1, 2, 3, 4, 5, 6, 7, 8, and 9, it is said to be in **standard form.** The position of each digit in the number determines the digit's **place value.** The diagram below shows a **place-value chart** naming the first twelve place values. The number 37,462 is in standard form and has been entered in the chart.

In the number 37,462, the position of the digit 3 determines that its place value is ten-thousands.

When a number is written in standard form, each group of digits separated by a comma is called a **period.** The number 3,786,451,294 has four periods. The period names are shown in red in the place-value chart above.

To write a number in words, start from the left. Name the number in each period. Then write the period name in place of the comma.

3,786,451,294 is read "three billion seven hundred eighty-six million four hundred fifty-one thousand two hundred ninety-four."

To write a whole number in standard form, write the number named in each period, and replace each period name with a comma.

Four million sixty-two thousand five hundred eighty-four is written 4,062,584 The zero is used as a place holder for the hundred-thousands' place.

Example 5 Write 25,478,083 in words.

Solution Twenty-five million four hundred seventy-eight thousand eighty-three

Example 6 Write 36,462,075 in words.

Your solution

Example 7 Write three hundred three thousand three in standard form.

Solution 303,003

Example 8 Write four hundred fifty-two thousand seven in standard form.

Your solution

Solutions on p. A7

Objective C To write whole numbers in expanded form

The whole number 26,429 can be written in **expanded form** as

20,000 + 6000 + 400 + 20 + 9

The place-value chart can be used to find the expanded form of a number.

Section 1.1 / Introduction to Whole Numbers

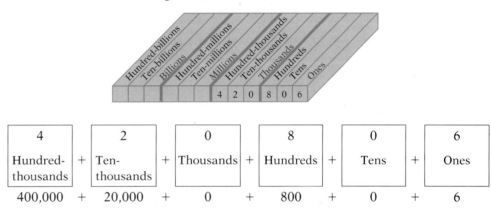

Write the number 420,806 in expanded form.

Notice the effect of having zeros in the number.

| 4 Hundred-thousands | + | 2 Ten-thousands | + | 0 Thousands | + | 8 Hundreds | + | 0 Tens | + | 6 Ones |
| 400,000 | + | 20,000 | + | 0 | + | 800 | + | 0 | + | 6 |

or simply 400,000 + 20,000 + 800 + 6

Example 9 Write 23,859 in expanded form.

Solution 20,000 + 3000 + 800 + 50 + 9

Example 10 Write 68,281 in expanded form.

Your solution

Example 11 Write 709,542 in expanded form.

Solution 700,000 + 9000 + 500 + 40 + 2

Example 12 Write 109,207 in expanded form.

Your solution

Solutions on p. A7

Objective D   **To round a whole number to a given place value**

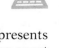

When the distance to the moon is given as 240,000 miles, the number represents an approximation to the true distance. Giving an approximate value for an exact number is called **rounding**. A number is always rounded to a given place value.

37 is closer to 40 than it is to 30. 37 rounded to the nearest ten is 40.

3673 rounded to the nearest ten is 3670. 3673 rounded to the nearest hundred is 3700.

A whole number is rounded to a given place value without using the number line by looking at the first digit to the right of the given place value.

If the digit to the right of the given place value is less than 5, that digit and all digits to the right are replaced by zeros.

Round 13,834 to the nearest hundred.

┌─ Given place value
13,834
 └── 3 < 5

13,834 rounded to the nearest hundred is 13,800.

If the digit to the right of the given place value is greater than or equal to 5, increase the digit in the given place value by 1, and replace all other digits to the right by zeros.

Round 386,217 to the nearest ten-thousand.

┌─ Given place value
386,217
  └── 6 > 5

386,217 rounded to the nearest ten-thousand is 390,000.

Example 13 Round 525,453 to the nearest ten-thousand.

Solution
┌─ Given place value
525,453
  └── 5 = 5

525,453 rounded to the nearest ten-thousand is 530,000.

Example 14 Round 368,492 to the nearest ten-thousand.

Your solution

Example 15 Round 1972 to the nearest hundred.

Solution
┌─ Given place value
1972
 └── 7 > 5

1972 rounded to the nearest hundred is 2000.

Example 16 Round 3962 to the nearest hundred.

Your solution

Solutions on p. A7

1.1 EXERCISES

▶ **Objective A**

Graph the number on the number line.

1. 3

2. 5

3. 9

4. 0

Place the correct symbol, < or >, between the two numbers.

5. 37   49   6. 58   21   7. 101   87

8. 16   5   9. 245 158   10. 2701   2071

11. 0   45   12. 107 0   13. 815   928

14. 2400 24,000   15. 7003 7020   16. 36,010   36,001

▶ **Objective B**

Write the number in words.

17. 805   18. 609   19. 485   20. 576

21. 2675   22. 3790   23. 42,928   24. 58,473

25. 356,943   26. 498,512   27. 3,697,483   28. 6,842,715

Write the number in standard form.

29. Eighty-five

30. Three hundred fifty-seven

31. Three thousand four hundred fifty-six

32. Sixty-three thousand seven hundred eighty

33. Six hundred nine thousand nine hundred forty-eight

34. Seven million twenty-four thousand seven hundred nine

35. Four million three thousand two

36. Six million five thousand eight

▶ Objective C

Write the number in expanded form.

37. 5287

38. 6295

39. 58,943

40. 453,921

41. 217,586

42. 4053

43. 50,943

44. 80,492

45. 200,583

46. 301,809

47. 403,705

48. 3,000,642

▶ Objective D

Round the number to the given place value.

49. 926 Tens

50. 845 Tens

51. 1439 Hundreds

52. 3973 Hundreds

53. 7238 Thousands

54. 7609 Thousands

55. 43,607 Thousands

56. 52,715 Thousands

57. 647,989 Ten-thousands

58. 253,678 Ten-thousands

SECTION 1.2 Addition of Whole Numbers

Objective A To add whole numbers without carrying

Addition is the process of finding the total of two or more numbers.

By counting, we see that the total of $3 and $4 is $7.

$3 + $4 = $7
Addend   Addend   Sum

Addition can be illustrated on the number line by using arrows to represent the addends. The size or magnitude of a number can be represented on the number line by an arrow.

The number 3 can be represented anywhere on the number line by an arrow that is 3 units in length.

To add on the number line, place the arrows representing the addends head to tail, with the first arrow starting at zero. The sum is represented by an arrow starting at zero and stopping at the tip of the last arrow.

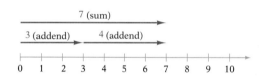

$3 + 4 = 7$

More than two numbers can be added on the number line.

$3 + 2 + 4 = 9$

Some special properties of addition that are used frequently are given below.

Addition Property of Zero
Zero added to a number does not change the number.

$4 + 0 = 4$
$0 + 7 = 7$

Commutative Property of Addition
Two numbers can be added in either order; the sum will be the same.

$4 + 8 = 8 + 4$
$12 = 12$

Associative Property of Addition
Grouping the addition in any order gives the same result. The parentheses are grouping symbols and have the meaning "do the operations inside the parentheses first."

$(4 + 2) + 3 = 4 + (2 + 3)$
$6 + 3 = 4 + 5$
$9 = 9$

The number line is not useful for adding large numbers. The basic addition facts for adding one digit to one digit are listed in the Appendix on page A2 for your review. Addition of larger numbers requires the repeated use of the basic addition facts.

To add large numbers, begin by arranging the numbers vertically, keeping the digits of the same place value in the same column.

Add: 321 + 6472

Add the digits in each column.

```
  THOUSANDS
   HUNDREDS
      TENS
      ONES
    3 2 1
 +6 4 7 2
  6 7 9 3
```

Example 1
Add: 4205 + 72,103

Solution
```
    4 2 0 5
 +7 2 1 0 3
  7 6 3 0 8
```

Example 2
Add: 3508 + 92,170

Your solution

Example 3
Add: 131 + 2135 + 3123 + 510

Solution
```
    1 3 1
    2 1 3 5
    3 1 2 3
 +    5 1 0
    5 8 9 9
```

Example 4
Add: 102 + 7351 + 1024 + 410

Your solution

Solutions on p. A7

Objective B To add whole numbers with carrying

In the previous section, numbers were added in which the sum of the numbers in any column did not exceed 9. When the sum of the digits in a column exceeds 9, the addition will involve "carrying."

Add: 487 + 369

Add the ones' column.
$7 + 9 = 16$ (1 ten + 6 ones).
Write the 6 in the ones' column and carry the 1 ten to the tens' column.

```
   HUNDREDS
      TENS
      ONES
      1
    4 8 7
  +3 6 9
        6
```

Add the tens' column.
$1 + 8 + 6 = 15$ (1 hundred + 5 tens).
Write the 5 in the tens' column and carry the 1 hundred to the hundreds' column.

```
    1 1
    4 8 7
  +3 6 9
      5 6
```

Add the hundreds' column.
1 + 4 + 3 = 8 (8 hundreds).
Write the 8 in the hundreds' column.

```
  1 1
  4 8 7
 +3 6 9
  8 5 6
```

Example 5   Add: 89 + 36 + 98

Solution
```
  2
  89
  36
 +98
 223
```

Example 6   Add: 95 + 88 + 67

Your solution

Example 7   Add: 41,395
                        4,327
                      497,625
                    + 32,991

Solution
```
 1 1 2  2 1
  41,395
   4,327
 497,625
+ 32,991
 576,338
```

Example 8   Add: 392
                        4,079
                       89,035
                     + 4,992

Your solution

Solutions on p. A7

ESTIMATION

Estimation and Calculators

At some places in the text you will be asked to use your calculator. Effective use of a calculator requires that you estimate the answer to the problem. This helps ensure that you have entered the numbers correctly and pressed the correct keys.

For example, if you use your calculator to find 22,347 + 5896 and the answer in the calculator's display is 131,757,912, you should realize that you have entered some part of the calculation incorrectly. In this case, you pressed ⨯ instead of +. By estimating the answer to a problem, you can help ensure the accuracy of your calculations. The symbol ≈ is used to denote **approximately equal.**

For example, to estimate the answer to
22,347 + 5896, round off each number
to the same place value. In this case we
will round to the nearest thousands. Then add.

```
 22,347 ≈   22,000
+ 5,896 ≈ +  6,000
            28,000
```

The sum 22,347 + 5896 is approximately 28,000. Knowing this, you would know that 131,757,912 is much too large and therefore incorrect.

To estimate the sum of two numbers, first round each whole number to the same place value and then add. Compare this answer with the calculator's answer.

Objective C To solve application problems

To solve an application problem, first read the problem carefully. The **Strategy** involves identifying the quantity to be found and planning the steps that are necessary to find that quantity. The **Solution** involves performing each operation stated in the Strategy and writing the answer.

Example 9
You had $1054 in your checking account before you made a deposit of $870. Find the amount in your checking account after the deposit.

Strategy

To find the amount in the checking account after the deposit, add the amount deposited ($870) to the amount in the checking account before the deposit ($1054).

Solution

$1054
+ 870
$1924

The amount in your checking account after the deposit was $1924.

Example 10
The odometer on your car read 23,972 miles before a 297-mile trip. What was the odometer reading after the trip?

Your strategy

Your solution

Example 11
Your paycheck shows deductions of $225 for savings, $98 for taxes, and $27 for insurance. Find the total of the three deductions.

Strategy

To find the total of the deductions, add the three amounts ($225, $98, and $27).

Solution

$225
  98
+ 27
$350

The total of the three deductions is $350.

Example 12
A homemaker has a monthly budget of $275 for food, $75 for car expenses, and $64 for entertainment. Find the total amount budgeted for the three items each month.

Your strategy

Your solution

Solutions on p. A7

1.2 EXERCISES

Objective A

Add.

1. 6
   +5

2. 7
   +8

3. 0
   +8

4. 3
   +0

5. 16
   + 3

6. 5
   +13

7. 40
   + 9

8. 8
   +70

9. 17
   +11

10. 25
   +63

11. 83
   +42

12. 63
   +94

13. 231
   +462

14. 845
   +152

15. 770
   +329

16. 542
   +920

17. 421
   +308

18. 202
   +725

19. 658
   +831

20. 842
   +936

21. 8092
   +6307

22. 5024
   +7902

23. 6721
   +9144

24. 6503
   +9285

25. 71,092
   +85,407

26. 62,304
   +94,093

27. 8004
   +6391

28. 6002
   +7297

Add.

29. 923,571
    +863,117

30. 580,236
    +516,643

31. 823,453
    +571,346

32. 673,914
    +716,055

33. 5 + 4 + 9

34. 3911 + 4045

35. 9144 + 7632

36. 6648 + 9221

37. 67,453 + 82,546

38. 73,581 + 66,418

39. 88,123 + 80,451

40. 44,765 + 82,033

41. 86,332 + 43,511

42. 97,531 + 91,346

▶ **Objective B**

Add.

43. 859
    +725

44. 637
    +829

45. 470
    +749

46. 427
    +690

47. 6016
    +5050

48. 4505
    +6009

49. 1897
    +3246

50. 8975
    +2146

51. 36,925
    +65,392

52. 56,772
    +51,239

53. 50,873
    +28,453

54. 34,872
    +46,079

Add.

55. 878
737
+189

56. 768
461
+669

57. 319
348
+912

58. 292
579
+315

59. 9409
3253
+7078

60. 8188
8020
+7104

61. 2038
2243
+3139

62. 4252
6882
+5235

63. 67,428
32,171
+20,971

64. 52,801
11,664
+89,638

65. 76,290
43,761
+87,402

66. 43,901
98,301
+67,943

67. 45,098
44,532
98,123
44,109
+77,310

68. 32,087
33,687
66,301
63,442
+91,842

69. 56,035
66,321
66,281
99,517
+55,109

70. 55,320
88,321
80,332
87,539
+10,321

Add.

71. $3 + 6 + 9 + 8 + 5$

72. $7 + 6 + 9 + 4 + 6$

73. $27 + 3 + 504$

74. $82 + 5 + 609$

75. $42 + 290 + 302$

76. $67 + 840 + 325$

77. $2709 + 658 + 10,935$

78. $8707 + 216 + 90,714$

Add.

79. 20,958 + 3218 + 42

80. 80,973 + 5168 + 29

81. 392 + 37 + 10,924 + 621

82. 694 + 62 + 70,129 + 217

83. 294 + 1029 + 7935 + 65

84. 692 + 2107 + 3196 + 92

85. 97 + 7234 + 69,532 + 276

86. 87 + 1698 + 27,317 + 727

87. 62 + 329 + 8954 + 1072

88. 87 + 946 + 6571 + 2103

89. 654 + 7293 + 237 + 33

90. 994 + 91,764 + 872 + 65

Estimate by rounding to the nearest hundreds. Then use your calculator to add.

91. 1234 + 9780 + 6740

92. 919 + 3642 + 8796

93. 241 + 569 + 390 + 1672

94. 107 + 984 + 1035 + 2904

Estimate by rounding to the nearest thousands. Then use your calculator to add.

95.   32,461
     9,844
  +59,407

96.   29,036
   22,904
  + 7,903

97.   25,432
   62,941
  +70,390

98.   66,541
   29,365
  +98,742

Estimate by rounding to the nearest ten-thousands. Then use your calculator to add.

99. 67,421
82,984
66,361
10,792
+34,037

100. 21,896
4,235
62,544
21,892
+ 1,334

101. 281,421
9,874
34,394
526,398
+ 94,631

102. 542,698
97,327
7,235
73,667
+173,201

Estimate by rounding to the nearest millions. Then use your calculator to add.

103. 28,627,052
983,073
+ 3,081,496

104. 1,792,085
29,919,301
+ 3,406,882

105. 12,377,491
3,409,723
7,928,026
+10,705,682

106. 46,751,070
6,095,832
280,011
+ 1,563,897

▶ Objective C *Application Problems*

107. The attendance at the Friday night rock concert was 2114, and the attendance at the Saturday night concert was 3678. Find the total attendance at the two concerts.

108. The attendance at the Saturday night professional baseball game was 35,946, and the attendance at the Sunday afternoon game was 27,429. Find the total attendance for the two games.

109. A quarterback threw passes for 42 yards in the first quarter, 117 yards in the second quarter, 66 yards in the third quarter, and 82 yards in the fourth quarter. Find the total number of yards gained by passing.

110. Your basketball team scored 17 points the first quarter, 26 points the second quarter, 14 points the third quarter, and 32 points the fourth quarter. Find the total number of points scored.

111. A computer manufacturer produced 3285 computers in January, 2714 computers in February, and 2182 computers in March. How many computers were manufactured during the 3 months?

112. An account executive received commissions of $3168 during October, $2986 during November, and $819 during December. Find the total commission received for the 3-month period.

113. There were 289,658 paid admissions for the first five games of a world series. There were 63,786 and 67,562 paid admissions for the sixth and seventh games of the series.
   a. Find the total paid attendance at the sixth and seventh games.
   b. Find the total paid attendance for the entire series.

114. A compact disc retail company sold 1578 compact discs during the first 9 months of the year. The company sold 98 compact discs in October, 167 in November, and 467 in December.
   a. How many compact discs were sold during the last 3 months of the year?
   b. What was the total number of compact discs sold during the year?

115. A student has $1348 in a checking account to be used for the fall semester. During the summer the student makes deposits of $518, $678, and $463.
   a. Find the total amount deposited.
   b. Find the new checking account balance.

116. The odometer on a moving van reads 68,692. The driver plans to drive 515 miles the first day, 492 miles the second day, and 278 miles the third day.
   a. How many miles will be driven during the 3 days?
   b. What will the odometer reading be at the end of the trip?

117. A clothing shop has the following monthly expenses: office supplies—$144, postage—$65, computer rental—$640, salaries—$2338, insurance—$212, rent—$1468, and utilities—$365. Find the total costs for the month.

118. An account executive keeps a record of miles driven for income tax purposes. The miles driven the previous 12 months were: 1544 miles, 2098 miles, 2536 miles, 1984 miles, 1690 miles, 2856 miles, 785 miles, 2348 miles, 1854 miles, 3054 miles, 2213 miles, and 1256 miles. Find the total number of miles driven during the year.

SECTION 1.3 Subtraction of Whole Numbers

Objective A To subtract whole numbers without borrowing

Subtraction is the process of finding the difference between two numbers.

By counting, we see that the difference between $8 and $5 is $3.

$8 − $5 = $3
Minuend Subtrahend Difference

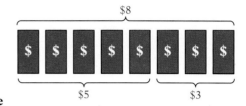

The difference 8 − 5 can be shown on the number line.

Note from the number line that addition and subtraction are related.

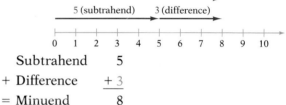

```
  Subtrahend      5
+ Difference    + 3
= Minuend        8
```

The fact that the sum of the subtrahend and difference equals the minuend can be used to check subtraction.

To subtract large numbers, begin by arranging the numbers vertically, keeping the digits of the same place value in the same column. Then subtract the digits in each column.

Subtract 8955 − 2432 and check.

```
           Check:   Subtrahend    2432
                  + Difference  + 6523
                  = Minuend       8955
```

Example 1 Subtract 6594 − 3271 and check.

Solution
```
  6594    Check:   3271
− 3271           + 3323
  3323             6594
```

Example 2 Subtract 8925 − 6413 and check.

Your solution

Example 3 Subtract 15,762 − 7541 and check.

Solution
```
  15,762   Check:   7,541
−  7,541          + 8,221
   8,221           15,762
```

Example 4 Subtract 17,504 − 9302 and check.

Your solution

Solutions on p. A8

Objective B To subtract whole numbers with borrowing

In all the subtraction problems in the previous objective, for each place value the lower digit was not larger than the upper digit. When the lower digit is larger than the upper digit, subtraction will involve "borrowing."

Subtract: 692 − 378

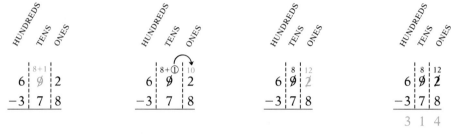

Because 8 > 2, borrowing is necessary.
9 tens = 8 tens + 1 ten.

Borrow 1 ten from the tens' column and write 10 in the ones' column.

Add the borrowed 10 to 2.

Subtract the digits in each column.

Subtract 751 − 234 and check.

```
    4 11              4 11                          1
  7 5 1             7 5 1         Check:         234
 −2 3 4            −2 3 4                       +517
                    5 1 7                        751
```

There may be more than one column in which borrowing is necessary.

Subtract 1234 − 485 and check.

```
    2 14           1 12 14        0 11 12 14                 1 1
  1 2 3 4         1 2 3 4         1 2 3 4       Check:      485
 −  4 8 5        −  4 8 5        −  4 8 5                  +749
          9            4 9            7 4 9               1234
```

Subtraction with a zero in the minuend involves repeated borrowing.

Subtract: 3904 − 1775

```
                                     9                       9
    8 10            8 10 14         8 10 14
  3 9 0 4         3 9 0 4         3 9 0 4
 −1 7 7 5        −1 7 7 5        −1 7 7 5
                                   2 1 2 9
```

5 > 4
There is a 0 in the tens' column. Borrow 1 hundred (= 10 tens) from the hundreds' column and write 10 in the tens' column.

Borrow 1 ten from the tens' column and add 10 to the 4 in the ones' column.

Subtract the digits in each column.

Example 5 Subtract 4392 − 678 and check.

Solution
```
   3 13 8 12
   4  3 9  2    Check:    678
 −    6 7 8            +3714
      3 7 1 4           4392
```

Example 6 Subtract 3481 − 865 and check.

Your solution

Example 7 Subtract 63,221 − 23,954 and check.

Solution
```
   5 12 11 11 11
   6  3, 2  2  1   Check:  23,954
 − 2  3, 9  5  4         +39,267
      3 9, 2  6  7        63,221
```

Example 8 Subtract 54,562 − 14,485 and check.

Your solution

Example 9 Subtract 46,005 − 32,167 and check.

Solution

```
   5 10
 4 6, 0 0 5
−3 2, 1 6 7
```
There are two zeros in the minuend. Borrow 1 thousand from the thousands' column and write 10 in the hundreds' column.

```
       9
   5  10 10
 4 6, 0  0  5
−3 2, 1  6  7
```
Borrow 1 hundred from the hundreds' column and write 10 in the tens' column.

```
       9  9
   5  10 10 15
 4 6, 0  0  5
−3 2, 1  6  7
 1 3, 8  3  8
```
Borrow 1 ten from the tens' column and add 10 to the 5 in the ones' column.

Check: 32,167
        +13,838
         46,005

Example 10 Subtract 64,003 − 54,936 and check.

Your solution

Solutions on p. A8

ESTIMATION

Estimating the Difference Between Two Whole Numbers

Estimate and then use your calculator to find 323,502 − 28,912.

To estimate the difference between two numbers, round each number to the same place value. In this case we will round to the nearest ten-thousands. Then subtract. The estimated answer is 290,000.

$$\begin{array}{r} 323{,}502 \approx 320{,}000 \\ -28{,}912 \approx -30{,}000 \\ \hline 290{,}000 \end{array}$$

Now use your calculator to find the exact result. The exact answer is 294,590.

323502 [−] 28912 [=] 294590

Objective C To solve application problems

Example 11

Your monthly food budget is $215. How much is left in the food budget after $148 is spent on groceries?

Strategy

To find the amount left in the food budget, subtract the amount already spent on groceries ($148) from the monthly food budget ($215).

Solution

$$\begin{array}{r} \$215 \\ -\ 148 \\ \hline \$\ 67 \end{array}$$

$67 is left in the monthly food budget.

Example 12

The down payment on a car costing $3250 is $675. Find the amount that remains to be paid.

Your strategy

Your solution

Example 13

You had a balance of $815 in your checking account. You then wrote checks in the amount of $112 for taxes, $57 for food, and $39 for shoes. What is your new checking account balance?

Strategy

To find your new checking account balance:

- Add to find the total of the three checks ($112 + $57 + $39).
- Subtract the total of the three checks from the old balance ($815).

Solution

$$\begin{array}{r} \$112 \\ 57 \\ +\ \ 39 \\ \hline \$208 \text{ total of checks} \end{array} \qquad \begin{array}{r} \$815 \\ -\ 208 \\ \hline \$607 \end{array}$$

Your new checking account balance is $607.

Example 14

Your total salary is $638. Deductions of $127 for taxes, $18 for insurance, and $35 for savings are taken from your pay. Find your take-home pay.

Your strategy

Your solution

Solutions on p. A8

1.3 EXERCISES

▶ Objective A

Subtract.

1. 9
   −5

2. 8
   −7

3. 8
   −4

4. 7
   −3

5. 10
   − 0

6. 7
   −0

7. 11
   − 4

8. 12
   − 8

9. 19
   − 8

10. 15
    − 6

11. 16
    − 7

12. 18
    − 9

13. 25
    − 3

14. 55
    − 4

15. 68
    − 8

16. 77
    − 3

17. 89
    −23

18. 68
    −43

19. 54
    −21

20. 88
    −57

21. 1202
    − 701

22. 1305
    − 404

23. 1763
    − 801

24. 1497
    − 706

25. 8974
    −3972

26. 2836
    −1711

27. 8976
    −7463

28. 9273
    −6142

29. 17 − 8

30. 15 − 7

31. 12 − 5

32. 93 − 71

33. 83 − 52

34. 77 − 36

35. 129 − 82

36. 132 − 61

37. 969 − 44

38. 1347 − 103

39. 4865 − 304

40. 1525 − 702

41. 9999 − 6794

42. 7806 − 3405

43. 8843 − 7621

44. 8713 − 6512

45. 5672 − 4361

46. 2900 − 1900

47. 7727 − 4326

48. 5988 − 1713

49. 2904 − 1403

Objective B

Subtract.

50. 71
    −18

51. 93
    −28

52. 47
    −18

53. 44
    −27

54. 71
    −67

55. 37
    −29

56. 50
    −27

57. 70
    −33

58. 993
    −537

59. 681
    −328

60. 250
    −192

61. 840
    −783

62. 768
    −194

63. 679
    −519

64. 770
    −395

65. 630
    −475

66. 893
    −874

67. 673
    −649

68. 508
    −299

69. 704
    −157

70. 67 − 39

71. 82 − 37

72. 80 − 32

73. 837 − 83

74. 642 − 87

75. 470 − 92

76. 810 − 380

77. 720 − 290

78. 674 − 337

79. 3526 − 387

80. 1712 − 289

81. 4350 − 729

82. 1702 − 948

83. 1607 − 869

84. 5933 − 3754

85. 7293 − 3748

86. 8143 − 2417

87. 7236 − 1978

88. 5714 − 2367

89. 8462 − 3575

90. 9407 − 2918

91. 3706 − 2957

92. 8605 − 7716

93. 8052 − 2709

94. 80,305 − 9176

95. 70,702 − 4239

96. 10,004 − 9306

97. 80,009 − 63,419

98. 70,618 − 41,213

99. 80,053 − 27,649

100. 70,700 − 21,076

101. 80,800 − 42,023

Subtract.

102. 95,432 − 87,857 **103.** 13,806 − 9439 **104.** 12,701 − 8624 **105.** 14,316 − 9427

106. 10,072 − 9514 **107.** 10,067 − 7428 **108.** 68,023 − 29,174

109. 3000
       −1785

110. 6000
       −2975

111. 7407
       −2359

112. 9703
       −2347

113. 2600
       −1972

114. 8400
       −3762

115. 9003
       −2471

116. 6004
       −2392

117. 8202
       −3916

118. 7050
       −4137

119. 7015
       −2973

120. 4207
       −1624

121. 7005
       −1796

122. 8003
       −2735

123. 20,005
       − 9,627

124. 80,004
       − 8,237

Estimate by rounding to the nearest ten-thousands. Then use your calculator to subtract.

125. 80,032
       −19,605

126. 90,765
       −60,928

127. 32,574
       −10,961

128. 96,430
       −59,762

129. 567,423
       −208,444

130. 300,712
       −198,714

Estimate by rounding to the nearest hundred-thousands. Then use your calculator to subtract.

131. 224,198
       − 96,441

132. 318,074
       − 64,908

133. 556,409
       − 99,642

134. 873,925
       − 28,744

135. 305,219
       −147,355

136. 608,123
       −238,107

Estimate by rounding to the nearest millions. Then use your calculator to subtract.

137.   3,050,092
        −1,967,325
	138. 9,000,765
	        −1,230,768
		139. 9,070,605
		        −3,492,763

140.   9,872,504
        −8,947,396
	141. 4,682,191
	        −3,056,822
		142. 9,947,812
		        −9,057,867

▶ **Objective C**   *Application Problems*

143. You have $304 in your checking account. If you write a check for $139, how much is left in your checking account?

144. You have $521 in your checking account. If you write a check for $247, how much is left in your checking account?

145. A tennis coach purchases a video camera that costs $1079 and makes a down payment of $180. Find the amount that remains to be paid.

146. A student purchases a used car that costs $5225 and makes a down payment of $550. Find the amount that remains to be paid.

147. The odometer of a rental car reads 2041 miles after a trip of 396 miles. What was the odometer reading at the start of the trip?

148. At the end of a vacation trip, the odometer of your car read 71,004 miles. If the odometer at the beginning of the trip read 68,238 miles, what was the length of the trip?

149. A steelworker receives a salary of $872 per week. Deductions from the check are $168 for taxes, $32 for union dues, $48 for social security, and $28 for insurance.
    a. Find the total amount deducted.
    b. Find the steelworker's take home pay.

150. A student had a bank balance of $6531. The student then wrote checks of $3248 for tuition, $116 for fees, and $279 for books.
    a. Find the total of the three checks.
    b. Find the new bank balance.

151. A family has a monthly budget of $1200. After $329 is spent for food, $128 for clothing, $82 for transportation, and $79 for entertainment, how much is left in the budget?

152. A wholesale distributor has 2405 computers in stock. After sales of 94 computers, 32 computers, and 116 computers, how many computers remain in stock?

153. A basketball team had an income of $13,425,320. Player salaries amounted to $9,125,320, and other expenses amounted to $3,215,860. Find the team's profit for the season.

154. The land area of Canada is 3,851,809 square miles and the land area of the United States is 3,675,633 square miles. How much larger than the United States is Canada?

SECTION 1.4 Multiplication of Whole Numbers

Objective A To multiply a number by a single digit

Six boxes of toasters are ordered. Each box contains eight toasters. How many toasters are ordered?

This problem can be worked by adding 6 eights.

$8 + 8 + 8 + 8 + 8 + 8 = 48$

This problem involves repeated addition of the same number and can be worked by a shorter process called **multiplication.**

Multiplication is the repeated addition of the same number.

$8 + 8 + 8 + 8 + 8 + 8 = 48$

or

The numbers that are multiplied are called **factors.** The answer is called the **product.**

$$6 \times 8 = 48$$
Factor Factor   Product

The product of 6×8 can be represented on the number line. The arrow representing the whole number 8 is repeated 6 times. The result is the arrow representing 48.

The times sign "×" is one symbol that is used to mean multiplication. Another common symbol used is a dot placed between the numbers.

$$7 \times 8 = 56 \qquad 7 \cdot 8 = 56$$

As with addition, there are some useful properties of multiplication.

Multiplication Property of Zero
The product of a number and zero is zero.

$0 \times 4 = 0$
$7 \times 0 = 0$

Multiplication Property of One
The product of a number and one is the number.

$1 \times 6 = 6$
$8 \times 1 = 8$

Commutative Property of Multiplication
Two numbers can be multiplied in either order. The product will be the same.

$4 \times 3 = 3 \times 4$
$12 = 12$

Associative Property of Multiplication
Grouping the multiplication in any order gives the same result. Do the multiplication inside the parentheses first.

$(4 \times 2) \times 3 = 4 \times (2 \times 3)$
$8 \times 3 = 4 \times 6$
$24 = 24$

The basic facts for multiplying one digit numbers are listed on page A2 in the Appendix for your review. Multiplication of larger numbers requires the repeated use of the basic multiplication facts.

Multiply: 37×4

Multiply 4×7. $4 \times 7 = 28$ (2 tens + 8 ones).
Write the 8 in the ones' column and
carry the 2 to the tens' column.

$$\begin{array}{r} \overset{2}{3}7 \\ \times4 \\ \hline 8 \end{array}$$

The 3 in 37 is 3 tens.
Multiply 4×3 tens. 4×3 tens = 12 tens
Add the carry digit. + 2 tens
              14 tens

$$\begin{array}{r} \overset{2}{3}7 \\ \times4 \\ \hline 148 \end{array}$$

Write the 14.

Example 1 Multiply: 64×8

Solution
$$\begin{array}{r} \overset{3}{6}4 \\ \times8 \\ \hline 512 \end{array}$$

Example 2 Multiply: 78×6

Your solution

Example 3 Multiply: 735×9

Solution
$$\begin{array}{r} \overset{3\,4}{7}35 \\ \times9 \\ \hline 6615 \end{array}$$

Example 4 Multiply: 648×7

Your solution

Solutions on p. A8

Objective B   To multiply larger whole numbers

Note the pattern when the following numbers are multiplied.

Multiply the nonzero part of the factors.

Now attach the same number of zeros to the product as the total number of zeros in the factors.

4×2
$4 \times 20 = 80$
    1 zero

4×2
$4 \times 200 = 800$
    2 zeros

4×2
$40 \times 200 = 8000$
    3 zeros

12×5
$12 \times 5000 = 60,000$
    3 zeros

Multiply: 47 × 23

Multiply by the ones' digit.	Multiply by the tens' digit.	Add.
47 ×23 --- 141 (= 47 × 3)	47 ×23 --- 141 940 (= 47 × 20) Writing the 0 is optional.	47 ×23 --- 141 94 --- 1081

The place-value chart illustrates the placement of the products.

Note the placement of the products when multiplying by a factor that contains a zero.

Multiply: 439 × 206

```
    439                When working the problem, usually        439
  ×206                 only one zero is written. Writing this  ×206
  ----                 zero ensures the proper placement of    ----
  2634                 the products.                           2634
   000    0 × 439                                              8780
  878                                                         -----
  ------                                                     90,434
  90,434
```

Example 5   Multiply: 829 × 603

Solution
```
    829
   ×603
   ----
   2487
  49740
  ------
 499,887
```

Example 6   Multiply: 756 × 305

Your solution

Solution on p. A8

ESTIMATION

Estimating the Product of Two Whole Numbers

Estimate and then use your calculator to find 3267 × 389.

To estimate a product, round each number so that all the digits are zero except the first digit. Then multiply. The estimated answer is 1,200,000.

```
  3267  ≈    3000
 × 389  ≈   × 400
            -------
           1,200,000
```

Now use your calculator to find the exact answer. The exact answer is 1,270,863.

Objective C To solve application problems

Example 7

An auto mechanic receives a salary of $525 each week. How much does the auto mechanic earn in 4 weeks?

Strategy

To find the mechanic's earnings for 4 weeks, multiply the weekly salary ($525) by the number of weeks (4).

Solution

$525
×   4
$2100

The mechanic earns $2100 in 4 weeks.

Example 8

A new car dealer receives a shipment of 37 cars each month. Find the number of cars the dealer will receive in 12 months.

Your strategy

Your solution

Example 9

A press operator earns $320 for working a 40-hour week. This week the press operator also worked 7 hours of overtime at $13 an hour. Find the press operator's total pay for the week.

Strategy

To find the press operator's income for the week:
- Find the overtime pay by multiplying the hours of overtime (7) by the overtime rate of pay ($13).
- Add the weekly salary ($320) to the overtime pay.

Solution

$13 $320
×   7   + 91
$91 overtime pay $411

The press operator earned $411 this week.

Example 10

The manager of a clothing store can buy 80 men's suits for $4800. Each sports jacket will cost the store $23. The manager orders 80 men's suits and 25 sports jackets. What is the total cost of the order?

Your strategy

Your solution

Solutions on p. A9

1.4 EXERCISES

▶ **Objective A**

Multiply.

1. 3 ×4	2. 2 ×8	3. 5 ×7	4. 6 ×4	5. 5 ×5
6. 7 ×7	7. 0 ×7	8. 8 ×0	9. 8 ×9	10. 7 ×6
11. 7 ×3	12. 2 ×9	13. 5 ×6	14. 7 ×4	15. 6 ×9
16. 8 ×5	17. 82 × 4	18. 79 × 3	19. 55 × 4	20. 72 × 2
21. 89 × 7	22. 96 × 5	23. 79 × 9	24. 67 × 7	25. 45 × 9
26. 66 × 3	27. 70 × 4	28. 67 × 5	29. 127 × 9	30. 623 × 4
31. 802 × 5	32. 607 × 9	33. 300 × 5	34. 600 × 7	35. 906 × 8
36. 703 × 9	37. 127 × 5	38. 632 × 3	39. 559 × 4	40. 632 × 8
41. 780 × 7	42. 690 × 5	43. 465 × 4	44. 382 × 7	45. 367 × 3

Multiply.

46. 524 × 4	**47.** 337 × 5	**48.** 841 × 6	**49.** 6709 × 7	**50.** 3608 × 5
51. 8568 × 7	**52.** 5495 × 4	**53.** 4780 × 4	**54.** 3690 × 5	**55.** 9895 × 2
56. 4697 × 3	**57.** 48,253 × 3	**58.** 69,587 × 8	**59.** 80,057 × 7	**60.** 90,032 × 7
61. 93,000 × 7	**62.** 87,000 × 9	**63.** 93,257 × 6	**64.** 88,524 × 7	

▶ **Objective B**

Multiply.

65. 16 ×21	**66.** 18 ×24	**67.** 35 ×26	**68.** 27 ×72	
69. 20 ×32	**70.** 30 ×47	**71.** 40 ×20	**72.** 30 ×70	
73. 39 ×46	**74.** 37 ×25	**75.** 67 ×23	**76.** 95 ×33	
77. 693 × 91	**78.** 581 × 72	**79.** 419 × 80	**80.** 727 × 60	
81. 8279 × 46	**82.** 9577 × 35	**83.** 6938 × 78	**84.** 8875 × 67	
85. 7035 × 57	**86.** 6702 × 48	**87.** 3009 × 35	**88.** 6003 × 57	

Multiply.

89.	3300 × 20	**90.**	5400 × 60	**91.**	3987 × 29	**92.**	4765 × 37
93.	809 ×530	**94.**	607 ×460	**95.**	800 ×325	**96.**	700 ×274
97.	987 ×349	**98.**	688 ×674	**99.**	312 ×134	**100.**	423 ×427
101.	379 ×500	**102.**	684 ×700	**103.**	985 ×408	**104.**	758 ×209
105.	3407 × 309	**106.**	5207 × 902	**107.**	4258 × 986	**108.**	6327 × 876
109.	8800 × 140	**110.**	9900 × 320	**111.**	2675 × 487	**112.**	3985 × 364

Estimate and then use your calculator to multiply.

113.	8745 × 63	**114.**	4732 × 93	**115.**	39,246 × 29	**116.**	64,409 × 67
117.	2937 × 206	**118.**	8941 × 726	**119.**	3097 ×1025	**120.**	6379 ×2936
121.	32,508 × 591	**122.**	62,504 × 923	**123.**	81,405 × 902	**124.**	66,735 × 844

Objective C Application Problems

125. A plane flying from Los Angeles to Boston uses 865 gallons of jet fuel each hour. How many gallons of jet fuel were used on a 5-hour flight?

126. A compact car averages 43 miles on 1 gallon of gas. How many miles could such a car travel on 13 gallons of gas?

127. A machine at a bottling company can fill and cap 4200 bottles of a soft drink in 1 hour. How many bottles of soft drink can the machine fill and cap in 40 hours?

128. A buyer for a department store purchased 225 suits at $64 each. What is the total cost for the 225 suits?

129. A computer programmer buys a car by making payments of $234 each month for 48 months. Find the total cost of the car.

130. An investor receives a check of $187 each month from a tax-free municipal bond fund. How much will the investor receive over a 12-month period?

131. A manufacturer of video cameras has 4347 video cameras in stock. The company can manufacture 16 video cameras each hour.
    a. How many video cameras can be manufactured in 12 hours?
    b. How many video cameras will be in stock 12 hours from now, assuming none are sold?

132. A gasoline storage tank contains 75,000 gallons of gasoline. When a valve is opened, a pipeline can deliver 37 gallons of gasoline each minute to the tank.
    a. How many gallons of gasoline are delivered to the tank in 28 minutes?
    b. How many gallons are in the tank after 28 minutes?

133. The tickets to a Christmas play were $4 for adults and $1 for children. 172 adult tickets and 136 child tickets were sold. Find the total income from the sale of the tickets.

134. A service station attendant earns $286 for working a 40-hour week. Last week the attendant worked an additional 8 hours at $13 an hour. Find the attendant's total pay for last week's work.

135. A computer graphing screen has 640 rows of pixels, and there are 480 pixels in a row. Find the total number of pixels on the screen.

136. A computer graphing screen has 720 rows of pixels, and there are 348 pixels in a row. Find the total number of pixels on the screen.

137. An oil company with 98,460,000 shares is bought by another company. The price per share is $88. How much was paid for the oil company?

138. An earth satellite travels 27,560 miles in making 1 revolution. The satellite makes 16 revolutions in 1 day. How many miles does the satellite travel in 30 days?

SECTION 1.5   Division of Whole Numbers

Objective A   To divide by a single digit with no remainder in the quotient

Division is used to separate objects into equal groups.

A store manager wants to distribute 24 new objects equally on 4 shelves. From the diagram, we see that the manager would place 6 objects on each shelf.

The manager's division problem can be written:

Notice that the quotient multiplied by the divisor equals the dividend.

6
$4\overline{)24}$ because | 6 | × | 4 | = | 24 |
                               | Quotient | | Divisor | | Dividend |

6
$9\overline{)54}$ because   6   ×   9   =   54

5
$8\overline{)40}$ because   5   ×   8   =   40

Here are some important quotients and the properties of 0 in division:

Important Quotients

Any whole number, except zero, divided by itself is 1.
$\quad 8\overline{)8}^{\,1} \quad 14\overline{)14}^{\,1} \quad 10\overline{)10}^{\,1}$

Any whole number divided by 1 is the whole number.
$\quad 1\overline{)9}^{\,9} \quad 1\overline{)27}^{\,27} \quad 1\overline{)10}^{\,10}$

Properties of Zero in Division

Zero divided by any other whole number is zero.
$\quad 7\overline{)0}^{\,0} \quad 13\overline{)0}^{\,0} \quad 10\overline{)0}^{\,0}$

Division by zero is not allowed.
$\quad 0\overline{)8}^{\,?}$ There is no number whose product with 0 is 8.

When the dividend is a larger whole number, the digits in the quotient are found in steps.

Divide $4\overline{)3192}$ and check.

$$
\begin{array}{r}
7 \\
4\overline{)3192} \\
-28 \\
\hline
39
\end{array}
$$
Think $4\overline{)31}$.
Subtract 7×4.
Bring down the 9.

$$
\begin{array}{r}
79 \\
4\overline{)3192} \\
-28 \\
\hline
39 \\
-36 \\
\hline
32
\end{array}
$$
Think $4\overline{)39}$.
Subtract 9×4.
Bring down the 2.

$$
\begin{array}{r}
798 \\
4\overline{)3192} \\
-28 \\
\hline
39 \\
-36 \\
\hline
32 \\
-32 \\
\hline
0
\end{array}
$$
Think $4\overline{)32}$.
Subtract 8×4.

Check:
$$
\begin{array}{r}
798 \\
\times 4 \\
\hline
3192
\end{array}
$$

The place-value chart can be used to show why this method works.

$$
\begin{array}{r}
\text{HUNDREDS · TENS · ONES} \\
7\ 9\ 8 \\
4\overline{)3\ 1\ 9\ 2} \\
-2\ 8\ 0\ 0 \\
\hline
3\ 9\ 2 \\
-3\ 6\ 0 \\
\hline
3\ 2 \\
-3\ 2 \\
\hline
0
\end{array}
$$

7 hundreds × 4

9 tens × 4

8 ones × 4

Division is also expressed by using the symbols " ÷ " or "—", which are read "divided by."

54 divided by 9 equals 6.
54 ÷ 9 equals 6.
$\dfrac{54}{9}$ equals 6.

Section 1.5 / Division of Whole Numbers   37

Example 1   Divide 7)56 and check.

Solution
$$\frac{8}{7\overline{)56}}$$

Check: 8 × 7 = 56

Example 2   Divide 9)63 and check.

Your solution

Example 3   Divide 2808 ÷ 8 and check.

Solution
```
      351
   8) 2808
     -24
      40
     -40
       08
      - 8
        0
```

Check: 351 × 8 = 2808

Example 4   Divide 4077 ÷ 9 and check.

Your solution

Example 5   Divide 7)2856 and check.

Solution
```
       408
    7) 2856
      -28
       05         Think 7)5.
      - 0         Place 0 in quotient.
       ___        Subtract 0 × 7.
        56        Bring down the 6.
       -56
         0
```

Check: 408 × 7 = 2856

Example 6   Divide 9)6345 and check.

Your solution

Solutions on p. A9

Objective B — To divide by a single digit with a remainder in the quotient

Occasionally it is not possible to separate objects into a whole number of equal groups.

A warehouse clerk must place 14 objects into 3 boxes. From the diagram, we see that the clerk would place 4 objects in each box and have 2 objects left over. The 2 is called the **remainder**.

The clerk's division problem could be written:

$$
\begin{array}{r}
\text{Number in each box} \\
\text{Quotient} \\
\text{Number of boxes} \to 3\overline{)14} \leftarrow \text{Total number of objects} \\
\text{Divisor} \quad\quad -12 \quad\quad \text{Dividend} \\
\hline
2 \leftarrow \text{Number left over} \\
\text{Remainder}
\end{array}
$$

The answer to a division problem with a remainder is frequently written:

$$3\overline{)14}\ \ 4\ r2$$

Notice that $\boxed{\underset{\text{Quotient}}{4} \times \underset{\text{Divisor}}{3}} + \boxed{\underset{\text{Remainder}}{2}} = \boxed{\underset{\text{Dividend}}{14}}$

Example 7 Divide $4\overline{)2522}$ and check.

Solution
$$
\begin{array}{r}
630\ r2 \\
4\overline{)2522} \\
-24 \\
\hline
12 \\
-12 \\
\hline
02 \\
-\ 0 \\
\hline
2
\end{array}
$$
Think $4\overline{)2}$.
Place 0 in quotient.
Subtract 0×4.

Check: $(630 \times 4) + 2 =$
$2520\ \ +2 = 2522$

Example 8 Divide $6\overline{)5225}$ and check.

Your solution

Example 9 Divide $9\overline{)27{,}438}$ and check.

Solution
$$
\begin{array}{r}
3{,}048\ r6 \\
9\overline{)27{,}438} \\
-27 \\
\hline
0\ 4 \\
-\ 0 \\
\hline
43 \\
-36 \\
\hline
78 \\
-72 \\
\hline
6
\end{array}
$$
Think $9\overline{)4}$.
Subtract 0×4.

Check: $(3048 \times 9) + 6 =$
$27{,}432\ \ + 6 = 27{,}438$

Example 10 Divide $7\overline{)21{,}409}$ and check.

Your solution

Solutions on p. A9

Objective C To divide by larger whole numbers

When the divisor has more than one digit, estimate at each step by using the first digit of the divisor. If that product is too large, lower the guess by 1 and try again.

Divide 34)1598 and check.

$$\begin{array}{r} 5 \\ 34 \overline{)1598} \\ -170 \end{array}$$ Think 3)15.
Subtract 5 × 34.

170 is too large. Lower guess by 1 and try again.

$$\begin{array}{r} 4 \\ 34 \overline{)1598} \\ -136 \\ \hline 238 \end{array}$$ Subtract 4 × 34.

$$\begin{array}{r} 47 \\ 34 \overline{)1598} \\ -136 \\ \hline 238 \\ -238 \\ \hline 0 \end{array}$$ Think 3)23.
Subtract 7 × 34.

Check:
$$\begin{array}{r} 47 \\ \times 34 \\ \hline 188 \\ 141 \\ \hline 1598 \end{array}$$

Example 11 Divide 7077 ÷ 34 and check.

Solution
$$\begin{array}{r} 208 \text{ r}5 \\ 34 \overline{)7077} \\ -68 \\ \hline 27 \\ -0 \\ \hline 277 \\ -272 \\ \hline 5 \end{array}$$

Think 34)27.
Place 0 in the quotient.
Subtract 0 × 34.

Check: (208 × 34) + 5 =
7072 + 5 = 7077

Example 12 Divide 4578 ÷ 42 and check.

Your solution

Solution on p. A9

Example 13 Divide $21{,}312 \div 56$ and check.

Solution

$$
\begin{array}{r}
380\ \text{r}32 \\
56\overline{)21{,}312} \\
-16\ 8 \\
\hline
4\ 51 \\
-4\ 48 \\
\hline
32 \\
-\ \ 0 \\
\hline
32
\end{array}
$$

Think $5\overline{)21}$.
4×56 is too large. Try 3.

Check: $(380 \times 56) + 32 =$
$21{,}280\ \ + 32 = 21{,}312$

Example 14 Divide $18{,}359 \div 39$ and check.

Your solution

Example 15 Divide $427\overline{)24{,}782}$ and check.

Solution

$$
\begin{array}{r}
58\ \text{r}16 \\
427\overline{)24{,}782} \\
-21\ 35 \\
\hline
3\ 432 \\
-3\ 416 \\
\hline
16
\end{array}
$$

Check: $(58 \times 427) + 16 =$
$24{,}766\ \ + 16 = 24{,}782$

Example 16 Divide $534\overline{)33{,}219}$ and check.

Your solution

Example 17 Divide $386\overline{)206{,}149}$ and check.

Solution

$$
\begin{array}{r}
534\ \text{r}25 \\
386\overline{)206{,}149} \\
-193\ 0 \\
\hline
13\ 14 \\
-11\ 58 \\
\hline
1\ 569 \\
-1\ 544 \\
\hline
25
\end{array}
$$

Check: $(534 \times 386) + 25 =$
$206{,}124\ \ + 25 = 206{,}149$

Example 18 Divide $515\overline{)216{,}848}$ and check.

Your solution

Solutions on p. A10

ESTIMATION

Estimating the Quotient of Two Whole Numbers

Estimate and then use your calculator to find 36,936 ÷ 54.

To estimate a quotient, round each number so that all the digits are zero except the first digit. Then divide.

36,936 ÷ 54 ≈
40,000 ÷ 50 = 800

The estimated answer is 800.

Now use your calculator to find the exact answer.

36 936 ÷ 54 = 684

The exact answer is 684.

Objective D To solve application problems

Example 19

A wheat farmer ships 192,600 bushels of wheat in 9 railroad cars. Find the amount of wheat shipped in each car.

Strategy

To find the amount of wheat shipped in each car, divide the number of bushels (192,600) by the number of cars (9).

Solution

```
       21,400
    9)192,600
     -18
       12
      - 9
        36
       -36
         0
```

Each car carried 21,400 bushels of wheat.

Example 20

A retail tire store can stock 270 tires on 15 shelves. How many tires can be stored on each shelf?

Your strategy

Your solution

Solution on p. A10

Example 21

The car you are buying costs $11,216. A down payment of $2000 is required. The remaining balance is paid in 48 equal monthly payments. What is the monthly car payment?

Strategy

To find the monthly payment:
- Find the remaining balance by subtracting the down payment ($2000) from the total cost of the car ($11,216).
- Divide the remaining balance by the number of equal monthly payments (48).

Solution

```
  $11,216                $ 192
 -  2,000             48)$9216
  $ 9,216 remaining    −48
          balance      ‾‾‾
                        441
                       −432
                       ‾‾‾
                         96
                        −96
                       ‾‾‾
                          0
```

The monthly payment is $192.

Example 22

A soft-drink manufacturer produces 12,600 cans of soft drink each hour. Cans are packed 24 to a case. How many cases of soft drink are produced in 8 hours?

Your strategy

Your solution

Solution on p. A10

1.5 EXERCISES

▶ Objective A

Divide.

1. $4\overline{)8}$
2. $3\overline{)9}$
3. $6\overline{)36}$
4. $9\overline{)81}$
5. $5\overline{)25}$

6. $7\overline{)49}$
7. $5\overline{)80}$
8. $6\overline{)96}$
9. $6\overline{)480}$
10. $9\overline{)630}$

11. $4\overline{)840}$
12. $3\overline{)690}$
13. $7\overline{)308}$
14. $7\overline{)203}$
15. $7\overline{)1442}$

16. $9\overline{)6327}$
17. $4\overline{)2120}$
18. $8\overline{)7280}$
19. $9\overline{)8118}$
20. $8\overline{)7264}$

21. $9\overline{)3510}$
22. $7\overline{)6020}$
23. $5\overline{)5280}$
24. $6\overline{)7218}$

25. $6\overline{)15{,}642}$
26. $6\overline{)19{,}254}$
27. $4\overline{)39{,}200}$
28. $3\overline{)16{,}200}$

29. $3\overline{)64{,}680}$
30. $4\overline{)50{,}760}$
31. $6\overline{)21{,}480}$
32. $5\overline{)18{,}050}$

▶ Objective B

Divide.

33. $4\overline{)9}$
34. $2\overline{)7}$
35. $5\overline{)27}$
36. $9\overline{)88}$
37. $3\overline{)40}$

38. $6\overline{)97}$
39. $8\overline{)83}$
40. $5\overline{)54}$
41. $7\overline{)632}$
42. $4\overline{)363}$

Divide.

43. 4)921 44. 7)845 45. 8)1635 46. 5)1548 47. 7)9432

48. 6)6352 49. 9)7004 50. 7)6001 51. 6)2350 52. 8)6370

53. 3)7252 54. 4)6538 55. 9)4157 56. 7)6172

57. 7)8124 58. 3)5162 59. 5)3542 60. 8)3274

61. 4)15,300 62. 7)43,500 63. 8)72,354 64. 5)43,542

▶ **Objective C**

Divide.

65. 27)96 66. 44)82 67. 42)87 68. 67)93

69. 41)897 70. 32)693 71. 23)784 72. 25)772

73. 74)600 74. 92)500 75. 70)329 76. 50)467

77. 36)7225 78. 44)8821 79. 19)3859 80. 32)9697

81. 23)4257 82. 62)9143 83. 92)6234 84. 87)4911

Divide.

85. 14)8422
86. 42)4497
87. 26)5209
88. 37)7403

89. 88)3127
90. 92)6177
91. 33)8943
92. 27)4765

93. 22)98,654
94. 77)83,629
95. 64)38,912
96. 78)31,434

97. 206)3097
98. 504)6504
99. 654)1217
100. 546)2344

101. 169)8542
102. 456)7723
103. 223)8927
104. 467)9344

Estimate and then use your calculator to divide.

105. 76)389,804
106. 53)117,925
107. 29)637,072
108. 67)738,072

109. 38)934,648
110. 34)906,304
111. 309)876,324
112. 642)323,568

113. 209)632,016
114. 614)332,174
115. 179)5,734,444
116. 374)7,712,254

117. 492)40,836
118. 609)56,028
119. 396)26,928
120. 234)69,030

Objective D   Application Problems

121. Four service organizations collected $548,000 to promote and provide community services. What amount did each organization receive if the money was divided evenly?

122. An insurance agent used 128 gallons of gas in traveling 3456 miles. Find the number of miles traveled on each gallon of gas.

123. A computer can store 2,211,840 bytes of information on six disks. How many bytes of information can be stored on one disk?

124. A flight engineer has an annual income of $41,496. Find the engineer's monthly income.

125. A computer analyst doing consulting work received $5376 for working 168 hours on a project. Find the hourly rate the consultant charged.

126. The total cost of a living room set, including finance charges, is $2856. This amount is to be repaid in 24 equal payments. Find the amount of each payment.

127. A family makes a down payment of $1560 on a car costing $10,536.
    a. What is the remaining balance to be paid?
    b. The balance is to be paid in 48 equal monthly payments. Find the monthly payment.

128. A farmer harvested 39,544 heads of lettuce from one plot of land and 38,456 from a second plot. The lettuce was packed in boxes with 24 heads in each box.
    a. What is the total number of heads of lettuce harvested from both plots?
    b. How many boxes were needed to pack the lettuce from the two plots?

129. An organization was hired to raise funds for five charities. Of the amount collected, $35,000 was used to pay expenses. The remaining amount of the $352,765 collected was divided equally among the five charities. What amount did each charity receive?

130. A manufacturer produces and boxes 240 computer keyboards each hour. Four keyboards are put in each box for shipment. How many boxes of computer keyboards can be produced in 12 hours?

131. A large corporation has a life and medical insurance plan that costs the corporation $158 per month for each employee. The monthly cost of the insurance plan is $666,128. How many employees participate in the insurance plan?

132. A limited partnership consisting of 2750 investors sold a piece of land for $5,959,250. Find the amount each investor received.

SECTION 1.6 Exponential Notation and the Order of Operations Agreement

Objective A   To simplify expressions containing exponents

Repeated multiplication of the same factor can be written two ways.

$3 \cdot 3 \cdot 3 \cdot 3 \cdot 3$ or 3^5 ← **exponent**

The exponent indicates how many times the factor occurs in the multiplication. The expression 3^5 is in **exponential notation.**

It is important to be able to read numbers written in exponential notation.

$6 = 6^1$ read "six to the first power" or just "six." Usually the exponent 1 is not written.
$6 \cdot 6 = 6^2$ read "six squared" or "six to the second power."
$6 \cdot 6 \cdot 6 = 6^3$ read "six cubed" or "six to the third power."
$6 \cdot 6 \cdot 6 \cdot 6 = 6^4$ read "six to the fourth power."
$6 \cdot 6 \cdot 6 \cdot 6 \cdot 6 = 6^5$ read "six to the fifth power."

Each place value in the place-value chart can be expressed as a power of 10.

$$
\begin{aligned}
\text{Ten} &= 10 = 10 = 10^1 \\
\text{Hundred} &= 100 = 10 \cdot 10 = 10^2 \\
\text{Thousand} &= 1000 = 10 \cdot 10 \cdot 10 = 10^3 \\
\text{Ten-Thousand} &= 10{,}000 = 10 \cdot 10 \cdot 10 \cdot 10 = 10^4 \\
\text{Hundred-Thousand} &= 100{,}000 = 10 \cdot 10 \cdot 10 \cdot 10 \cdot 10 = 10^5 \\
\text{Million} &= 1{,}000{,}000 = 10 \cdot 10 \cdot 10 \cdot 10 \cdot 10 \cdot 10 = 10^6
\end{aligned}
$$

To simplify a numerical expression containing exponents, write each factor as many times as indicated by the exponent and carry out the indicated multiplication.

$4^3 = 4 \cdot 4 \cdot 4 = 64$
$2^2 \cdot 3^4 = (2 \cdot 2) \cdot (3 \cdot 3 \cdot 3 \cdot 3) = 4 \cdot 81 = 324$

Example 1   Write $3 \cdot 3 \cdot 3 \cdot 5 \cdot 5$ in exponential notation.

Solution $3 \cdot 3 \cdot 3 \cdot 5 \cdot 5 = 3^3 \cdot 5^2$

Example 2   Write $2 \cdot 2 \cdot 2 \cdot 2 \cdot 3 \cdot 3 \cdot 3$ in exponential notation.

Your solution

Example 3   Write $10 \cdot 10 \cdot 10 \cdot 10$ as a power of 10.

Solution $10 \cdot 10 \cdot 10 \cdot 10 = 10^4$

Example 4   Write $10 \cdot 10 \cdot 10 \cdot 10 \cdot 10 \cdot 10$ as a power of 10.

Your solution

Example 5   Simplify $3^2 \cdot 5^3$

Solution $3^2 \cdot 5^3 = (3 \cdot 3) \cdot (5 \cdot 5 \cdot 5)$
$= 9 \cdot 125 = 1125$

Example 6   Simplify $2^3 \cdot 5^2$.

Your solution

Solutions on p. A11

Objective B — To use the Order of Operations Agreement to simplify expressions

More than one operation may occur in a numerical expression. The answer may be different, depending on the order in which the operations are performed. For example, consider $3 + 4 \times 5$.

Multiply first, then add.

$$3 + \underbrace{4 \times 5}$$
$$\underbrace{3 + 20}$$
$$23$$

Add first, then multiply.

$$\underbrace{3 + 4} \times 5$$
$$\underbrace{7 \times 5}$$
$$35$$

An Order of Operations Agreement is used so that only one answer is possible.

The Order of Operations Agreement:

Step 1 Do all operations inside parentheses.
Step 2 Simplify any number expressions containing exponents.
Step 3 Do multiplication and division as they occur from left to right.
Step 4 Do addition and subtraction as they occur from left to right.

Simplify $3 \times (2 + 1) - 2^2 + 4 \div 2$ by using the Order of Operations Agreement.

$3 \times \underbrace{(2 + 1)} - 2^2 + 4 \div 2$ 1. Perform operations in parentheses.

$3 \times 3 - \underbrace{2^2} + 4 \div 2$ 2. Simplify expressions with exponents.

$\underbrace{3 \times 3} - 4 + 4 \div 2$ 3. Do multiplications and divisions as they occur from left to right.
$9 - 4 + \underbrace{4 \div 2}$

$\underbrace{9 - 4} + 2$ 4. Do additions and subtractions as they occur from left to right.
$\underbrace{5 + 2}$
7

One or more of the above steps may not be needed to simplify an expression. In that case, proceed to the next step in the Order of Operations Agreement.

Simplify $5 + 8 \div 2$.   No parentheses or exponents. Proceed to step 3 of the Agreement.

$5 + \underbrace{8 \div 2}$ 3. Do multiplication or division.

$\underbrace{5 + 4}$ 4. Do addition or subtraction.
9

Example 7 Simplify $16 \div (8 - 4) \cdot 9 - 5^2$.

Solution
$16 \div (8 - 4) \cdot 9 - 5^2$
$16 \div 4 \cdot 9 - 5^2$
$16 \div 4 \cdot 9 - 25$
$4 \cdot 9 - 25$
$36 - 25$
11

Example 8 Simplify $5 \cdot (8 - 4) \div 4 - 2$.

Your solution

Solution on p. A11

1.6 EXERCISES

▶ **Objective A**

Write the number in exponential notation.

1. $2 \cdot 2 \cdot 2$
2. $7 \cdot 7 \cdot 7 \cdot 7 \cdot 7$
3. $6 \cdot 6 \cdot 6 \cdot 7 \cdot 7 \cdot 7 \cdot 7$
4. $3 \cdot 3 \cdot 3 \cdot 5 \cdot 5 \cdot 5 \cdot 5$

5. $6 \cdot 6 \cdot 6 \cdot 11 \cdot 11 \cdot 11 \cdot 11 \cdot 11$
6. $2 \cdot 2 \cdot 2 \cdot 2 \cdot 2 \cdot 2 \cdot 3 \cdot 3 \cdot 3$

7. $3 \cdot 10 \cdot 10 \cdot 10 \cdot 10$
8. $7 \cdot 10 \cdot 10 \cdot 10 \cdot 10 \cdot 10 \cdot 10 \cdot 10$

9. $2 \cdot 2 \cdot 3 \cdot 3 \cdot 3 \cdot 5 \cdot 5 \cdot 5 \cdot 5$
10. $7 \cdot 7 \cdot 11 \cdot 11 \cdot 11 \cdot 13 \cdot 13 \cdot 13 \cdot 13$

11. $2 \cdot 3 \cdot 3 \cdot 7 \cdot 7 \cdot 11$
12. $5 \cdot 7 \cdot 7 \cdot 9 \cdot 9 \cdot 11$

13. $2 \cdot 2 \cdot 7 \cdot 7 \cdot 7 \cdot 7 \cdot 11 \cdot 11 \cdot 11 \cdot 11$
14. $3 \cdot 3 \cdot 7 \cdot 7 \cdot 7 \cdot 7 \cdot 17 \cdot 17 \cdot 17 \cdot 17$

Simplify.

15. 2^3
16. 2^6
17. $2^4 \cdot 5^2$
18. $2^6 \cdot 3^2$
19. $3^2 \cdot 10^2$

20. $2^3 \cdot 10^4$
21. $6^2 \cdot 3^3$
22. $4^3 \cdot 5^2$
23. $5 \cdot 2^3 \cdot 3$
24. $6 \cdot 3^2 \cdot 4$

25. $2^2 \cdot 3^2 \cdot 10$
26. $3^2 \cdot 5^2 \cdot 10$
27. $0^2 \cdot 4^3$
28. $6^2 \cdot 0^3$
29. $3^2 \cdot 10^4$

30. $5^3 \cdot 10^3$
31. $2^2 \cdot 3^3 \cdot 5$
32. $5^2 \cdot 7^3 \cdot 2$
33. $2 \cdot 3^4 \cdot 5^2$
34. $6 \cdot 2^6 \cdot 7^2$

35. $5^2 \cdot 3^2 \cdot 7^2$
36. $4^2 \cdot 9^2 \cdot 6^2$
37. $3^4 \cdot 2^6 \cdot 5$
38. $4^3 \cdot 6^3 \cdot 7$
39. $4^2 \cdot 3^3 \cdot 10^4$

40. $5^2 \cdot 2^3 \cdot 10^3$
41. $6^2 \cdot 4^4 \cdot 10$

Objective B

Simply by using the Order of Operations Agreement.

42. $4 - 2 + 3$
43. $6 - 3 + 2$
44. $6 \div 3 + 2$
45. $8 \div 4 + 8$

46. $6 \cdot 3 + 5$
47. $5 \cdot 9 + 2$
48. $3^2 - 4$
49. $5^2 - 17$

50. $4 \cdot (5 - 3) + 2$
51. $3 + (4 + 2) \div 3$
52. $5 + (8 + 4) \div 6$
53. $8 - 2^2 + 4$

54. $16 \cdot (3 + 2) \div 10$
55. $12 \cdot (1 + 5) \div 12$
56. $10 - 2^3 + 4$
57. $5 \cdot 3^2 + 8$

58. $16 + 4 \cdot 3^2$
59. $12 + 4 \cdot 2^3$
60. $16 + (8 - 3) \cdot 2$
61. $7 + (9 - 5) \cdot 3$

62. $2^2 + 3 \cdot (6 - 2)$
63. $3^3 + 5 \cdot (8 - 6)$
64. $2^2 \cdot 3^2 + 2 \cdot 3$
65. $4 \cdot 6 + 3^2 \cdot 4^2$

66. $16 - 2 \cdot 4$
67. $12 + 3 \cdot 5$
68. $3 \cdot (6 - 2) + 4$

69. $5 \cdot (8 - 4) - 6$
70. $8 - (8 - 2) \div 3$
71. $12 - (12 - 4) \div 4$

72. $8 + 2 - 3 \cdot 2 \div 3$
73. $10 + 1 - 5 \cdot 2 \div 5$
74. $3 \cdot (4 + 2) \div 6$

Some calculators use the Order of Operations Agreement, others do not. Try the following problems with your calculator to see if it uses the Order of Operations Agreement.

75. $2^2 + 8 \div 2$
76. $5^2 + 12 \div 6$
77. $4 \cdot 3 \div 6 + 5$

78. $8 \cdot 2 \div 4 - 3$
79. $3 \cdot (4 - 1) + 3^2$
80. $5 \cdot (7 - 2) - 2^3$

SECTION 1.7 Prime Numbers and Factoring

Objective A   To factor numbers

Whole number factors of a number divide that number evenly (there is no remainder).

1, 2, 3, and 6 are whole number factors of 6 because they divide 6 evenly.

$$\begin{array}{cccc} 6 & 3 & 2 & 1 \\ 1\overline{)6} & 2\overline{)6} & 3\overline{)6} & 6\overline{)6} \end{array}$$

Notice that both the divisor and the quotient are factors of the dividend.

To find the factors of a number, try dividing the number by 1, 2, 3, 4, 5, . . . Those numbers that divide the number evenly are its factors. Continue this process until the factors start to repeat.

Find all the factors of 42.
42 ÷ 1 = 42   1 and 42 are factors
42 ÷ 2 = 21   2 and 21 are factors
42 ÷ 3 = 14   3 and 14 are factors
42 ÷ 4   Will not divide evenly
42 ÷ 5   Will not divide evenly
42 ÷ 6 = 7   6 and 7 are factors ⎫ Factors are repeating; all the factors
42 ÷ 7 = 6   7 and 6 are factors ⎭ of 42 have been found.

1, 2, 3, 6, 7, 14, 21, and 42 are factors of 42.

The following rules are helpful in finding the factors of a number.

2 is a factor of a number if the last digit of the number is 0, 2, 4, 6, or 8.

436 ends in 6; therefore, 2 is a factor of 436. (436 ÷ 2 = 218)

3 is a factor of a number if the sum of the digits of the number is divisible by 3.

The sum of the digits of 489 is 4 + 8 + 9 = 21. 21 is divisible by 3. Therefore, 3 is a factor of 489. (489 ÷ 3 = 163)

5 is a factor of a number if the last digit of the number is 0 or 5.

520 ends in 0; therefore, 5 is a factor of 520. (520 ÷ 5 = 104)

Example 1   Find all the factors of 30.

Solution   30 ÷ 1 = 30
30 ÷ 2 = 15
30 ÷ 3 = 10
30 ÷ 4   Will not divide evenly
30 ÷ 5 = 6
30 ÷ 6 = 5

1, 2, 3, 5, 6, 10, 15 and 30 are factors of 30.

Example 2   Find all the factors of 40.

Your solution

Solution on p. A11

Objective B To find the prime factorization of a number

A number is a **prime** number if its only whole number factors are 1 and itself. 7 is prime because its only factors are 1 and 7. If a number is not prime, it is called a **composite** number. Because 6 has factors of 2 and 3, 6 is a composite number. The number 1 is not considered a prime number; therefore it is not included in the following list of prime numbers less than 50.

$$2, 3, 5, 7, 11, 13, 17, 19, 23, 29, 31, 37, 41, 43, 47$$

The **prime factorization** of a number is the expression of the number as a product of its prime factors. To find the prime factors of 60, begin with the smallest prime number as a trial divisor and continue with prime numbers as trial divisors until the final quotient is prime. A "T-diagram" is sometimes used as a short version of the procedure.

$$\begin{array}{r} & & & 5 \text{ Prime} \\ & & 15 & 3\overline{)15} \\ & 30 & 2\overline{)30} & 2\overline{)30} \\ 2\overline{)60} & 2\overline{)60} & 2\overline{)60} \end{array}$$

T-diagram Version

$$\begin{array}{r|l} 60 & \\ 2 & 30 \\ 2 & 15 \\ 3 & 5 \text{ Prime} \end{array}$$

The prime factorization of 60 is $2 \cdot 2 \cdot 3 \cdot 5$.

Finding the prime factorization of larger numbers can be more difficult. Try each prime number as a trial divisor. Stop when the square of the trial divisor is greater than the number being factored.

Find the prime factorization of 106.

$$\begin{array}{r} 53 \\ 2\overline{)106} \end{array}$$
53 cannot be divided evenly by 2, 3, 5, 7, or 11. Prime numbers greater than 11 need not be tested because 11^2 is greater than 53.

The prime factorization of 106 is $2 \cdot 53$.

Example 3 Find the prime factorization of 84.

Solution
$$\begin{array}{r} 7 \\ 3\overline{)21} \\ 2\overline{)42} \\ 2\overline{)84} \end{array}$$

$84 = 2 \cdot 2 \cdot 3 \cdot 7$

Example 4 Find the prime factorization of 44.

Your solution

Example 5 Find the prime factorization of 201.

Solution
$$\begin{array}{r} 67 \\ 3\overline{)201} \end{array}$$
Try only 2, 3, 5, 7, and 11 because $11^2 > 67$.

$201 = 3 \cdot 67$

Example 6 Find the prime factorization of 177.

Your solution

Solutions on p. A11

1.7 EXERCISES

▶ Objective A

Find all the factors of the number.

1. 4
2. 6
3. 10
4. 20
5. 5

6. 7
7. 12
8. 9
9. 8
10. 16

11. 13
12. 17
13. 18
14. 24
15. 25

16. 56
17. 36
18. 45
19. 28
20. 32

21. 29
22. 33
23. 22
24. 26
25. 44

26. 52
27. 49
28. 82
29. 37
30. 79

31. 57
32. 69
33. 48
34. 64
35. 87

36. 95
37. 46
38. 54
39. 50
40. 75

41. 66
42. 77
43. 80
44. 100
45. 84

46. 96
47. 85
48. 90
49. 101
50. 123

Find all the factors of the number.

51. 120 **52.** 150 **53.** 125 **54.** 140 **55.** 132

56. 144 **57.** 135 **58.** 180 **59.** 189 **60.** 168

▶ **Objective B**

Find the prime factorization.

61. 6   **62.** 14   **63.** 17   **64.** 83   **65.** 16

66. 24   **67.** 12   **68.** 27   **69.** 9   **70.** 15

71. 36   **72.** 40   **73.** 19   **74.** 37   **75.** 50

76. 90   **77.** 65   **78.** 115 **79.** 80   **80.** 87

81. 18   **82.** 26   **83.** 28   **84.** 49   **85.** 29

86. 31   **87.** 42   **88.** 62   **89.** 81   **90.** 51

91. 22   **92.** 39   **93.** 101 **94.** 89   **95.** 70

96. 66   **97.** 86   **98.** 74   **99.** 95   **100.** 105

101. 67 **102.** 78 **103.** 55 **104.** 46 **105.** 122

106. 130 **107.** 118 **108.** 102 **109.** 150 **110.** 125

111. 120 **112.** 144 **113.** 160 **114.** 175 **115.** 343

116. 216 **117.** 400 **118.** 625 **119.** 225 **120.** 180

Calculators and Computers

Order of Operations Does your calculator use the Order of Operations Agreement? To find out, try this problem:

$$2 + 4 \cdot 7$$

If your answer is 30, then the calculator uses the Order of Operations Agreement. If your answer is 42, it does not use that agreement.

Even if your calculator does not use the Order of Operations Agreement, you can still correctly evaluate numerical expressions. The parentheses keys, $[\,(\,]$ and $[\,)\,]$, are used for this purpose.

Remember that $2 + 4 \cdot 7$ means $2 + (4 \cdot 7)$ because the multiplication is completed before addition. To evaluate this expression, enter the following:

Enter	Display	Comments
2	2.	
$[+]$	2.	
$[\,(\,]$	(	Not all scientific calculators display the "(" when it is entered.
4	4.	
$[\times]$	4.	
7	7.	
$[\,)\,]$	28.	The product inside the parentheses.
$[=]$	30.	The answer to $2 + (4 \times 7)$.

When using your calculator to evaluate numerical expressions, insert parentheses around multiplications or divisions. This has the effect of forcing the calculator to do the operations in the order you want rather than the order the calculator wants.

Evaluate $3 \cdot (15 - 2 \cdot 3) - 36 \div 3$.

Press the keys shown in the box.

$$3 \times (15 - (2 \times 3)) - (36 \div 3) =$$

The answer 15 should be in the display.

Notice that parentheses were used around $15 - 2 \cdot 3$ and around $2 \cdot 3$. The parentheses are important to ensure that multiplication is performed before additions or subtractions from the innermost grouping symbols to the outermost grouping symbols.

Chapter Summary

Key Words The *whole numbers* are 0, 1, 2, 3, 4, 5, 6, 7, 8, 9, 10, . . .

The symbol for *"is less than"* is $<$.

The symbol for *"is greater than"* is $>$.

The position of a digit in a number determines the digit's *place value*.

Giving an approximate value for an exact number is called *"rounding."*

An expression of the form 4^2 is in *exponential notation*, where 4 is the *base* and 2 is the *exponent*.

A number is *prime* if its only whole number factors are 1 and itself.

The *prime factorization* of a number is the expression of the number as a product of its prime factors.

Essential Rules

The Addition Property of Zero
Zero added to a number does not change the number.
$5 + 0 = 0 + 5 = 5$

The Commutative Property of Addition
Two numbers can be added in either order.
$6 + 5 = 5 + 6 = 11$

The Associative Property of Addition
Grouping the addition in any order gives the same result.
$(2 + 3) + 5 = 2 + (3 + 5) = 10$

The Multiplication Property of Zero
The product of a number and zero is zero.
$0 \times 3 = 3 \times 0 = 0$

The Multiplication Property of 1
The product of a number and 1 is the number.
$1 \times 8 = 8 \times 1 = 8$

The Commutative Property of Multiplication
Two numbers can be multiplied in any order.
$5 \times 2 = 2 \times 5 = 10$

The Associative Property of Multiplication
Grouping multiplication in any order gives the same result.
$(3 \times 2) \times 5 = 3 \times (2 \times 5) = 30$

The Properties of Zero in Division
Zero divided by any other number is zero. Division by zero is not allowed.

Order of Operations Agreement

Step 1 Perform operations inside grouping symbols.
Step 2 Simplify expressions with exponents.
Step 3 Do multiplications and divisions as they occur from left to right.
Step 4 Do additions and subtractions as they occur from left to right.

Chapter Review

SECTION 1.1

1. Place the correct symbol, < or >, between the two numbers. 101   87

2. Write 276,057 in words.

3. Write two million eleven thousand forty-four in standard form.

4. Write 10,327 in expanded form.

SECTION 1.2

5. Add: 298
         461
        +322

6. Add: 5894
          6301
        + 298

7. An insurance account executive receives commissions of $723, $544, $812, and $488 during a 4-week period. Find the total income from commissions for the 4 weeks.

8. You had a balance of $516 in your checking account before making deposits of $88 and $213. Find the total amount deposited, and determine your new checking balance.

SECTION 1.3

9. Subtract: 4926
             −3177

10. Subtract: 10,134
              − 4,725

11. After a trip of 972 miles, the odometer of your car read 52,031 miles. What was the odometer reading at the beginning of the trip?

12. Your monthly budget for household expenses is $600. After you spend $137 for food and $67 for utilities, how much is left in your budget?

SECTION 1.4

13. Multiply: 843
              × 27

14. Multiply: 2,019
              ×   307

15. You have a car payment of $123 per month. What is the total of the car payments over a 12-month period?

16. A sales assistant earns $240 for working a 40-hour week. Last week the assistant worked an additional 12 hours at $12 an hour. Find the assistant's total pay for last week's work.

SECTION 1.5

17. Divide: $7 \overline{)14{,}945}$

18. Divide: $84 \overline{)109{,}763}$

19. A sales executive drove a car 351 miles on 13 gallons of gas. Find the number of miles driven per gallon of gasoline.

20. A car is purchased for $8940, with a down payment of $1500. The balance is paid in 48 equal monthly payments. Find the monthly car payment.

SECTION 1.6

21. Write $2 \cdot 2 \cdot 2 \cdot 2 \cdot 5 \cdot 5 \cdot 5$ in exponential notation.

22. Write $5 \cdot 5 \cdot 7 \cdot 7 \cdot 7 \cdot 7 \cdot 7$ in exponential notation.

23. Simplify $3 \cdot 2^3 \cdot 5^2$.

24. Simplify $2^3 - 3 \cdot 2$.

25. Simplify $3^2 + 2^2 \cdot (5 - 3)$

26. Simplify $8 \cdot (6 - 2) \div 4$.

SECTION 1.7

27. Find all the factors of 18.

28. Find all the factors of 30.

29. Find the prime factorization of 42.

30. Find the prime factorization of 72.

Chapter Test

1. Place the correct symbol, < or >, between the two numbers. 21   19

2. Write 207,068 in words.

3. Write one million two hundred four thousand six in standard form.

4. Write 906,378 in expanded form.

5. Round 74,965 to the nearest hundred.

6. Add: 25,492
        +71,306

7. Add: 89,756
           9,094
         +37,065

8. A family drives 425 miles the first day, 187 miles the second day, and 243 miles the third day of their vacation. The odometer read 47,626 miles at the start of the vacation.
   a. How many miles were driven during the 3 days?
   b. What is the odometer reading at the end of the 3 days?

9. Subtract: 17,495
              − 8,162

10. Subtract: 20,736
               − 9,854

11. A flight attendant purchases a car costing $16,405 and makes a down payment of $2450. What is the remaining balance to be paid?

12. Multiply: 90,763
              ×   8

13. Multiply: 9736
              × 704

14. An investor receives $237 each month from a corporate bond fund. How much will the investor receive over a 12-month period?

15. Divide: $8\overline{)5624}$

16. Divide: $7\overline{)60,972}$

17. Divide: $97\overline{)108,764}$

18. A farmer harvested 48,290 pounds of lemons from one grove and 23,710 pounds of lemons from another grove. The lemons were packed in boxes with 24 pounds in each box. How many boxes were needed to pack the lemons?

19. Write $3 \cdot 3 \cdot 3 \cdot 7 \cdot 7$ in exponential notation.

20. Simplify $3^3 \cdot 4^2$.

21. Simplify $4^2 \cdot (4 - 2) \div 8 + 5$.

22. Simplify $16 \div 4 \times 2 - (6 - 5)^3$

23. Find all the factors of 20.

24. Find the prime factorization of 84.

25. Find the prime factorization of 102.

2 Fractions

OBJECTIVES

- To find the least common multiple (LCM)
- To find the greatest common factor (GCF)
- To write a fraction that represents part of a whole
- To write an improper fraction as a mixed number or a whole number, and a mixed number as an improper fraction
- To find equivalent fractions by raising to higher terms
- To write a fraction in simplest form
- To add fractions with the same denominator
- To add fractions with unlike denominators
- To add whole numbers, mixed numbers, and fractions
- To solve application problems
- To subtract fractions with the same denominator
- To subtract fractions with unlike denominators
- To subtract whole numbers, mixed numbers, and fractions
- To solve application problems
- To multiply fractions
- To multiply whole numbers, mixed numbers, and fractions
- To solve application problems
- To divide fractions
- To divide whole numbers, mixed numbers, and fractions
- To solve application problems
- To identify the order relation between two fractions
- To simplify expressions containing exponents
- To use the Order of Operations Agreement to simplify expressions

Egyptian Fractions

The Rhind papyrus is one of the earliest written accounts of mathematics.[1] In the papyrus, a scribe named Ahmes gives an early account of the concept of fractions. A portion of the Rhind papyrus is rendered below with its heiroglyphic transcription.

The early Egyptian primarily used unit fractions. This is a fraction in which the numerator is a 1. To write a fraction, a small oval was placed above a series of lines. The number of lines indicated the denominator. Some examples of these fractions are

$$\frac{\bigcirc}{||||} = \frac{1}{4} \qquad \sqsupset = \frac{1}{2}$$

In the first example, each line represents a 1. Because there are 4 lines, the fraction is $\frac{1}{4}$.

The second example is the special symbol that was used for the fraction $\frac{1}{2}$.

---

[1] Papyrus comes from the stem of a plant. The stem was dried and then pounded thin. The resulting material served as a primitive type of paper.

SECTION 2.1 The Least Common Multiple and Greatest Common Factor

Objective A — To find the least common multiple (LCM)

The **multiples** of a number are the products of that number and the numbers 1, 2, 3, 4, 5, . . .

$3 \times 1 = 3$
$3 \times 2 = 6$
$3 \times 3 = 9$
$3 \times 4 = 12$ The multiples of 3 are 3, 6, 9, 12, 15, . . .
$3 \times 5 = 15$
⋮

A number that is a multiple of two or more numbers is a **common multiple** of those numbers.

The multiples of 4 are: 4, 8, 12, 16, 20, 24, 28, 32, 36, . . .
The multiples of 6 are: 6, 12, 18, 24, 30, 36, 42, . . .
Some common multiples of 4 and 6 are: 12, 24, and 36.

The **least common multiple** (LCM) is the smallest common multiple of two or more numbers.

The least common multiple of 4 and 6 is 12.

Listing the multiples of each number is one way to find the LCM. Another way to find the LCM uses the prime factorization of each number.

To find the LCM of 450 and 600, find the prime factorization of each number and write the factorization of each number in a table. Circle the largest product in each column. The LCM is the product of the circled numbers.

	2	3	5
450 =	2	(3 · 3)	5 · 5
600 =	(2 · 2 · 2)	3	(5 · 5)

In the column headed by 5, the products are equal. Circle just one product.

The LCM is the product of the circled numbers.
The LCM = 2 · 2 · 2 · 3 · 3 · 5 · 5 = 1800.

Example 1   Find the LCM of 24, 36, and 50.

Solution

	2	3	5
24 =	(2 · 2 · 2)	3	
36 =	2 · 2	(3 · 3)	
50 =	2		(5 · 5)

The LCM =
2 · 2 · 2 · 3 · 3 · 5 · 5 = 1800.

Example 2   Find the LCM of 50, 84, and 135.

Your solution

Solution on p. A11

Objective B To find the greatest common factor (GCF)

Recall that a number that divides another number evenly is a factor of that number. 64 can be evenly divided by 1, 2, 4, 8, 16, 32, and 64. 1, 2, 4, 8, 16, 32, and 64 are factors of 64.

A number that is a factor of two or more numbers is a **common factor** of those numbers.

The factors of 30 are: 1, 2, 3, 5, 6, 10, 15, 30.
The factors of 105 are: 1, 3, 5, 7, 15, 21, 35, 105.
The common factors of 30 and 105 are: 1, 3, 5, and 15.

The **greatest common factor** (GCF) is the largest common factor of two or more numbers.

The greatest common factor of 30 and 105 is 15.

Listing the factors of each number is one way of finding the GCF. Another way to find the GCF uses the prime factorization of each number.

To find the GCF of 126 and 180, find the prime factorization of each number and write the factorization of each number in a table. Circle the smallest product in each column that does not have a blank. The GCF is the product of the circled numbers.

	2	3	5	7
126 =	②	③·3		7
180 =	2·2	3·3	5	

In the column headed by 3, the products are equal. Circle just one product.
Columns 5 and 7 have a blank, so 5 and 7 are not common factors of 126 and 180. Do not circle any number in these columns.

The GCF is the product of the circled numbers.
The GCF = 2 · 3 · 3 = 18.

Example 3 Find the GCF of 90, 168, and 420.

Solution

	2	3	5	7
90 =	②	3·3	5	
168 =	2·2·2	③		7
420 =	2·2	3	5	7

The GCF = 2 · 3 = 6.

Example 4 Find the GCF of 36, 60, and 72.

Your solution

Example 5 Find the GCF of 7, 12, and 20.

Solution

	2	3	5	7
7 =				7
12 =	2·2	3		
20 =	2·2		5	

Since no numbers are circled, the GCF = 1.

Example 6 Find the GCF of 11, 24, and 30.

Your solution

Solutions on p. A11

2.1 EXERCISES

▶ **Objective A**

Find the LCM.

1. 5, 8
2. 3, 6
3. 3, 8
4. 2, 5
5. 5, 6

6. 5, 7
7. 4, 6
8. 6, 8
9. 8, 12
10. 12, 16

11. 5, 12
12. 3, 16
13. 8, 14
14. 6, 18
15. 3, 9

16. 4, 10
17. 8, 32
18. 7, 21
19. 9, 36
20. 14, 42

21. 44, 60
22. 120, 160
23. 102, 184
24. 123, 234
25. 4, 8, 12

26. 5, 10, 15
27. 3, 5, 10
28. 2, 5, 8
29. 3, 8, 12
30. 5, 12, 18

31. 9, 36, 64
32. 18, 54, 63
33. 16, 30, 84
34. 9, 12, 15

35. 12, 18, 24
36. 30, 60, 80
37. 6, 9, 15
38. 8, 18, 24

39. 12, 15, 25
40. 13, 26, 39
41. 10, 15, 40
42. 12, 48, 72

Objective B

Find the GCF.

43. 3, 5   **44.** 5, 7   **45.** 6, 9   **46.** 18, 24   **47.** 15, 25

48. 14, 49   **49.** 25, 100   **50.** 16, 80   **51.** 32, 51   **52.** 21, 44

53. 12, 80   **54.** 8, 36   **55.** 16, 140   **56.** 12, 76   **57.** 8, 14

58. 24, 30   **59.** 48, 144   **60.** 44, 96   **61.** 18, 32   **62.** 16, 30

63. 40, 64   **64.** 24, 140   **65.** 60, 82   **66.** 40, 68   **67.** 2, 5, 7

68. 3, 5, 11   **69.** 6, 8, 10   **70.** 7, 14, 49   **71.** 6, 15, 36   **72.** 8, 12, 16

73. 10, 15, 20   **74.** 12, 18, 20   **75.** 24, 40, 72   **76.** 3, 17, 51

77. 17, 31, 81   **78.** 14, 42, 84   **79.** 25, 125, 625   **80.** 12, 68, 92

81. 28, 35, 70   **82.** 1, 49, 153   **83.** 32, 56, 72   **84.** 24, 36, 48

SECTION 2.2 Introduction to Fractions

Objective A   To write a fraction that represents part of a whole

A **fraction** can represent the number of equal parts of a whole.

The shaded portion of the circle is represented by the fraction $\frac{4}{7}$.

Four sevenths of the circle are shaded.

Each part of a fraction has a name.

Fraction bar $\rightarrow \frac{4}{7}$ $\leftarrow$ **Numerator**
$\phantom{\text{Fraction bar} \rightarrow \frac{4}{7}}$ $\leftarrow$ **Denominator**

A **proper fraction** is a fraction less than 1. The numerator of a proper fraction is smaller than the denominator. The shaded portion of the circle can be represented by the proper fraction $\frac{3}{4}$.

A **mixed number** is a number greater than 1 with a whole number part and a fractional part. The shaded portion of the circles can be represented by the mixed number $2\frac{1}{4}$.

An **improper fraction** is a fraction greater than or equal to 1. The numerator of an improper fraction is greater than or equal to the denominator. The shaded portion of the circles can be represented by the improper fraction $\frac{9}{4}$. The shaded portion of the square can be represented by $\frac{4}{4}$.

Example 1 Express the shaded portion of the circles as a mixed number.

Solution $3\frac{2}{5}$

Example 2 Express the shaded portion of the circles as a mixed number.

Your solution

Example 3 Express the shaded portion of the circles as an improper fraction.

Solution $\frac{17}{5}$

Example 4 Express the shaded portion of the circles as an improper fraction.

Your solution

Solutions on p. A11

Objective B To write an improper fraction as a mixed number or a whole number, and a mixed number as an improper fraction

Note from the diagram that the mixed number $2\frac{3}{5}$ and the improper fraction $\frac{13}{5}$ represent the shaded portion of the circles.

$$2\frac{3}{5} = \frac{13}{5}$$

An improper fraction can be written as a mixed number.

Write $\frac{13}{5}$ as a mixed number.

Divide the numerator by the denominator.

To write the fractional part of the mixed number, write the remainder over the divisor.

Write the answer.

$$\begin{array}{r} 2 \\ 5\overline{)13} \\ -10 \\ \hline 3 \end{array}$$

$$\begin{array}{r} 2\frac{3}{5} \\ 5\overline{)13} \\ -10 \\ \hline 3 \end{array}$$

$$\frac{13}{5} = 2\frac{3}{5}$$

To write a mixed number as an improper fraction, multiply the denominator of the fractional part by the whole number part. The sum of this product and the numerator of the fractional part is the numerator of the improper fraction. The denominator remains the same.

Write $7\frac{3}{8}$ as an improper fraction. $7\frac{3}{8} = \frac{(8 \times 7) + 3}{8} = \frac{56 + 3}{8} = \frac{59}{8}$ $7\frac{3}{8} = \frac{59}{8}$

Example 5 Write $\frac{21}{4}$ as a mixed number.

Solution
$$\begin{array}{r} 5 \\ 4\overline{)21} \\ -20 \\ \hline 1 \end{array} \qquad \frac{21}{4} = 5\frac{1}{4}$$

Example 6 Write $\frac{22}{5}$ as a mixed number.

Your solution

Example 7 Write $\frac{18}{6}$ as a whole number.

Solution
$$\begin{array}{r} 3 \\ 6\overline{)18} \\ -18 \\ \hline 0 \end{array} \quad \text{Note: the remainder is zero.} \quad \frac{18}{6} = 3$$

Example 8 Write $\frac{28}{7}$ as a whole number.

Your solution

Example 9 Write $21\frac{3}{4}$ as an improper fraction.

Solution $21\frac{3}{4} = \frac{84 + 3}{4} = \frac{87}{4}$

Example 10 Write $14\frac{5}{8}$ as an improper fraction.

Your solution

Solutions on p. A11

2.2 EXERCISES

▶ **Objective A**

Express the shaded portion of the circle as a fraction.

1.   2.

3.   4.

Express the shaded portion of the circles as a mixed number.

5.   6.

7.   8.

9.   10.

Express the shaded portion of the circles as an improper fraction.

11.   12.

13.   14.

15.   16.

Objective B

Write the improper fraction as a mixed number or whole number.

17. $\dfrac{11}{4}$ 18. $\dfrac{16}{3}$ 19. $\dfrac{20}{4}$ 20. $\dfrac{18}{9}$ 21. $\dfrac{9}{8}$ 22. $\dfrac{13}{4}$

23. $\dfrac{23}{10}$ 24. $\dfrac{29}{2}$ 25. $\dfrac{48}{16}$ 26. $\dfrac{51}{3}$ 27. $\dfrac{8}{7}$ 28. $\dfrac{16}{9}$

29. $\dfrac{7}{3}$ 30. $\dfrac{9}{5}$ 31. $\dfrac{16}{1}$ 32. $\dfrac{23}{1}$ 33. $\dfrac{17}{8}$ 34. $\dfrac{31}{16}$

35. $\dfrac{12}{5}$ 36. $\dfrac{19}{3}$ 37. $\dfrac{9}{9}$ 38. $\dfrac{40}{8}$ 39. $\dfrac{72}{8}$ 40. $\dfrac{3}{3}$

Write the mixed number as an improper fraction.

41. $2\dfrac{1}{3}$ 42. $4\dfrac{2}{3}$ 43. $6\dfrac{1}{2}$ 44. $8\dfrac{2}{3}$ 45. $6\dfrac{5}{6}$ 46. $7\dfrac{3}{8}$

47. $9\dfrac{1}{4}$ 48. $6\dfrac{1}{4}$ 49. $10\dfrac{1}{2}$ 50. $15\dfrac{1}{8}$ 51. $8\dfrac{1}{9}$ 52. $3\dfrac{5}{12}$

53. $5\dfrac{3}{11}$ 54. $3\dfrac{7}{9}$ 55. $2\dfrac{5}{8}$ 56. $12\dfrac{2}{3}$ 57. $1\dfrac{5}{8}$ 58. $5\dfrac{3}{7}$

59. $11\dfrac{1}{9}$ 60. $12\dfrac{3}{5}$ 61. $3\dfrac{3}{8}$ 62. $4\dfrac{5}{9}$ 63. $6\dfrac{7}{13}$ 64. $8\dfrac{5}{14}$

SECTION 2.3 Writing Equivalent Fractions

Objective A   To find equivalent fractions by raising to higher terms

Equal fractions with different denominators are called **equivalent fractions**.

$\frac{4}{6}$ is equivalent to $\frac{2}{3}$.

$\frac{2}{3}$ was rewritten as the equivalent fraction $\frac{4}{6}$.

Remember that the Multiplication Property of 1 stated that the product of a number and 1 is the number. This is true for fractions as well as whole numbers. This property can be used to write equivalent fractions.

$\frac{2}{3} \times 1 = \frac{2}{3} \times \frac{1}{1} = \frac{2 \cdot 1}{3 \cdot 1} = \frac{2}{3}$

$\frac{2}{3} \times 1 = \frac{2}{3} \times \boxed{\frac{2}{2}} = \frac{2 \cdot 2}{3 \cdot 2} = \frac{4}{6}$ $\frac{4}{6}$ is equivalent to $\frac{2}{3}$.

$\frac{2}{3} \times 1 = \frac{2}{3} \times \boxed{\frac{4}{4}} = \frac{2 \cdot 4}{3 \cdot 4} = \frac{8}{12}$ $\frac{8}{12}$ is equivalent to $\frac{2}{3}$.

$\frac{2}{3}$ was rewritten as the equivalent fractions $\frac{4}{6}$ and $\frac{8}{12}$.

Write a fraction that is equivalent to $\frac{5}{8}$ and has a denominator of 32.

Divide the larger denominator by the smaller. Multiply the numerator and denominator of the given fraction by the quotient (4).

$8\overline{)32}$ quotient 4

$\frac{5}{8} = \frac{5 \cdot 4}{8 \cdot 4} = \frac{20}{32}$

$\frac{20}{32}$ is equivalent to $\frac{5}{8}$.

Example 1 Write $\frac{2}{3}$ as an equivalent fraction that has a denominator of 42.

Solution $3\overline{)42}$ → 14 $\frac{2}{3} = \frac{2 \cdot 14}{3 \cdot 14} = \frac{28}{42}$

$\frac{28}{42}$ is equivalent to $\frac{2}{3}$.

Example 2 Write $\frac{3}{5}$ as an equivalent fraction that has a denominator of 45.

Your solution

Example 3 Write 4 as a fraction that has a denominator of 12.

Solution Write 4 as $\frac{4}{1}$.

$1\overline{)12}$ → 12 $4 = \frac{4 \cdot 12}{1 \cdot 12} = \frac{48}{12}$

$\frac{48}{12}$ is equivalent to 4.

Example 4 Write 6 as a fraction that has a denominator of 18.

Your solution

Solutions on p. A12

Objective B   To write a fraction in simplest form

A fraction is in **simplest form** when there are no common factors in the numerator and the denominator.

The fractions $\frac{4}{6}$ and $\frac{2}{3}$ are equivalent fractions.

$\frac{4}{6}$ has been written in simplest form as $\frac{2}{3}$.

The Multiplication Property of 1 can be used to write fractions in simplest form. Write the numerator and denominator of the given fraction as a product of factors. Write factors common to both the numerator and denominator as an improper fraction equivalent to 1.

$$\frac{4}{6} = \frac{2 \cdot 2}{2 \cdot 3} = \frac{2}{2} \cdot \frac{2}{3} = 1 \cdot \frac{2}{3} = \frac{2}{3}$$

The process of eliminating common factors is usually written as shown at the right.

$$\frac{4}{6} = \frac{\cancel{2} \cdot 2}{\cancel{2} \cdot 3} = \frac{2}{3}$$

To write a fraction in simplest form, eliminate the common factors.

$$\frac{18}{30} = \frac{\cancel{2} \cdot \cancel{3} \cdot 3}{\cancel{2} \cdot \cancel{3} \cdot 5} = \frac{3}{5}$$

An improper fraction should be changed to a mixed number.

$$\frac{22}{6} = \frac{\cancel{2} \cdot 11}{\cancel{2} \cdot 3} = \frac{11}{3} = 3\frac{2}{3}$$

Example 5   Write $\frac{15}{40}$ in simplest form.

Solution $\frac{15}{40} = \frac{3 \cdot \cancel{5}}{2 \cdot 2 \cdot 2 \cdot \cancel{5}} = \frac{3}{8}$

Example 6   Write $\frac{16}{24}$ in simplest form.

Your solution

Example 7   Write $\frac{6}{42}$ in simplest form.

Solution $\frac{6}{42} = \frac{\cancel{2} \cdot \cancel{3}}{\cancel{2} \cdot \cancel{3} \cdot 7} = \frac{1}{7}$

Example 8   Write $\frac{8}{56}$ in simplest form.

Your solution

Example 9   Write $\frac{8}{9}$ in simplest form.

Solution $\frac{8}{9} = \frac{2 \cdot 2 \cdot 2}{3 \cdot 3} = \frac{8}{9}$

$\frac{8}{9}$ is in simplest form because there are no common factors in the numerator and denominator.

Example 10   Write $\frac{15}{32}$ in simplest form.

Your solution

Example 11   Write $\frac{30}{12}$ in simplest form.

Solution $\frac{30}{12} = \frac{\cancel{2} \cdot \cancel{3} \cdot 5}{\cancel{2} \cdot 2 \cdot \cancel{3}} = \frac{5}{2} = 2\frac{1}{2}$

Example 12   Write $\frac{48}{36}$ in simplest form.

Your solution

Solutions on p. A12

2.3 EXERCISES

▶ Objective A

Write an equivalent fraction with the given denominator.

1. $\frac{1}{2} = \frac{}{10}$
2. $\frac{1}{4} = \frac{}{16}$
3. $\frac{3}{16} = \frac{}{48}$
4. $\frac{5}{9} = \frac{}{81}$
5. $\frac{3}{8} = \frac{}{32}$

6. $\frac{7}{11} = \frac{}{33}$
7. $\frac{3}{17} = \frac{}{51}$
8. $\frac{7}{10} = \frac{}{90}$
9. $\frac{3}{4} = \frac{}{16}$
10. $\frac{5}{8} = \frac{}{32}$

11. $3 = \frac{}{9}$
12. $5 = \frac{}{25}$
13. $\frac{1}{3} = \frac{}{60}$
14. $\frac{1}{16} = \frac{}{48}$
15. $\frac{11}{15} = \frac{}{60}$

16. $\frac{3}{50} = \frac{}{300}$
17. $\frac{2}{3} = \frac{}{18}$
18. $\frac{5}{9} = \frac{}{36}$
19. $\frac{5}{7} = \frac{}{49}$
20. $\frac{7}{8} = \frac{}{32}$

21. $\frac{5}{9} = \frac{}{18}$
22. $\frac{11}{12} = \frac{}{36}$
23. $7 = \frac{}{3}$
24. $9 = \frac{}{4}$
25. $\frac{7}{9} = \frac{}{45}$

26. $\frac{5}{6} = \frac{}{42}$
27. $\frac{15}{16} = \frac{}{64}$
28. $\frac{11}{18} = \frac{}{54}$
29. $\frac{3}{14} = \frac{}{98}$
30. $\frac{5}{6} = \frac{}{144}$

31. $\frac{5}{8} = \frac{}{48}$
32. $\frac{7}{12} = \frac{}{96}$
33. $\frac{5}{14} = \frac{}{42}$
34. $\frac{2}{3} = \frac{}{42}$
35. $\frac{17}{24} = \frac{}{144}$

36. $\frac{5}{13} = \frac{}{169}$
37. $\frac{3}{8} = \frac{}{408}$
38. $\frac{9}{16} = \frac{}{272}$
39. $\frac{17}{40} = \frac{}{800}$
40. $\frac{9}{25} = \frac{}{1000}$

41. $\frac{5}{12} = \frac{}{72}$
42. $\frac{9}{17} = \frac{}{119}$
43. $\frac{17}{30} = \frac{}{210}$
44. $\frac{11}{21} = \frac{}{189}$

Objective B

Write the fraction in simplest form.

45. $\dfrac{4}{12}$ 46. $\dfrac{8}{22}$ 47. $\dfrac{22}{44}$ 48. $\dfrac{2}{14}$ 49. $\dfrac{2}{12}$ 50. $\dfrac{6}{25}$

51. $\dfrac{50}{75}$ 52. $\dfrac{40}{36}$ 53. $\dfrac{12}{8}$ 54. $\dfrac{36}{9}$ 55. $\dfrac{0}{30}$ 56. $\dfrac{10}{10}$

57. $\dfrac{9}{22}$ 58. $\dfrac{14}{35}$ 59. $\dfrac{75}{25}$ 60. $\dfrac{8}{60}$ 61. $\dfrac{16}{84}$ 62. $\dfrac{14}{45}$

63. $\dfrac{20}{44}$ 64. $\dfrac{12}{35}$ 65. $\dfrac{8}{36}$ 66. $\dfrac{28}{44}$ 67. $\dfrac{12}{16}$ 68. $\dfrac{48}{35}$

69. $\dfrac{16}{12}$ 70. $\dfrac{24}{18}$ 71. $\dfrac{24}{40}$ 72. $\dfrac{44}{60}$ 73. $\dfrac{8}{88}$

74. $\dfrac{9}{90}$ 75. $\dfrac{144}{36}$ 76. $\dfrac{140}{297}$ 77. $\dfrac{48}{144}$ 78. $\dfrac{32}{120}$

79. $\dfrac{60}{100}$ 80. $\dfrac{33}{110}$ 81. $\dfrac{36}{16}$ 82. $\dfrac{80}{45}$ 83. $\dfrac{32}{160}$

84. $\dfrac{192}{640}$ 85. $\dfrac{19}{51}$ 86. $\dfrac{23}{146}$ 87. $\dfrac{81}{168}$ 88. $\dfrac{66}{220}$

SECTION 2.4 Addition of Fractions and Mixed Numbers

Objective A To add fractions with the same denominator

Fractions with the same denominator are added by adding the numerators and placing the sum over the common denominator. After adding, write the sum in simplest form.

Add: $\frac{2}{7} + \frac{4}{7}$

$\begin{array}{r} \frac{2}{7} \\ +\frac{4}{7} \\ \hline \frac{6}{7} \end{array}$ Add the numerators and place the sum over the common denominator.

$\frac{2}{7} + \frac{4}{7} = \frac{2+4}{7} = \frac{6}{7}$

Example 1 Add: $\frac{5}{12} + \frac{11}{12}$

Solution
$\begin{array}{r} \frac{5}{12} \\ +\frac{11}{12} \\ \hline \frac{16}{12} = \frac{4}{3} = 1\frac{1}{3} \end{array}$

Example 2 Add: $\frac{3}{8} + \frac{7}{8}$

Your solution

Solution on p. A12

Objective B To add fractions with unlike denominators

To add fractions with unlike denominators, first rewrite the fractions as equivalent fractions with a common denominator. The common denominator is the LCM of the denominators of the fractions.

Add: $\frac{1}{2} + \frac{1}{3}$

The common denominator is the LCM of 2 and 3. LCM = 6. The LCM of denominators is sometimes called the **least common denominator** (LCD).

Write equivalent fractions using the LCM.

$\frac{1}{2} = \frac{3}{6}$

$+\frac{1}{3} = \frac{2}{6}$

Add the fractions.

$\frac{1}{2} = \frac{3}{6}$

$+\frac{1}{3} = \frac{2}{6}$

$\overline{\frac{5}{6}}$

Example 3 Add: $\frac{3}{8} + \frac{7}{12}$

Solution
$$\frac{3}{8} = \frac{9}{24}$$ The LCM of 8 and 12 is 24.
$$+\frac{7}{12} = \frac{14}{24}$$
$$\overline{\frac{23}{24}}$$

Example 4 Add: $\frac{5}{12} + \frac{9}{16}$

Your solution

Example 5 Add: $\frac{5}{8} + \frac{7}{9}$

Solution
$$\frac{5}{8} = \frac{45}{72}$$ The LCM of 8 and 9 is 72.
$$+\frac{7}{9} = \frac{56}{72}$$
$$\overline{\frac{101}{72} = 1\frac{29}{72}}$$

Example 6 Add: $\frac{7}{8} + \frac{11}{15}$

Your solution

Example 7 Add: $\frac{2}{3} + \frac{3}{5} + \frac{5}{6}$

Solution
$$\frac{2}{3} = \frac{20}{30}$$ The LCM of 3, 5, and 6 is 30.
$$\frac{3}{5} = \frac{18}{30}$$
$$+\frac{5}{6} = \frac{25}{30}$$
$$\overline{\frac{63}{30} = 2\frac{3}{30} = 2\frac{1}{10}}$$

Example 8 Add: $\frac{3}{4} + \frac{4}{5} + \frac{5}{8}$

Your solution

Solutions on p. A12

Objective C To add whole numbers, mixed numbers, and fractions

The sum of a whole number and a fraction is a mixed number.

Add: $2 + \frac{2}{3}$

$$\boxed{2} + \frac{2}{3} = \boxed{\frac{6}{3}} + \frac{2}{3} = \frac{8}{3} = 2\frac{2}{3}$$

To add a whole number and a mixed number, write the fraction, then add the whole numbers.

Add: $7\frac{2}{5} + 4$

Write the fraction.

$$7\frac{2}{5}$$
$$+4$$
$$\overline{\frac{2}{5}}$$

Add the whole numbers.

$$7\frac{2}{5}$$
$$+4$$
$$\overline{11\frac{2}{5}}$$

Section 2.4 / Addition of Fractions and Mixed Numbers

To add two mixed numbers, add the fractional parts and then add the whole numbers. Remember to reduce the sum to simplest form.

Add: $5\frac{4}{9} + 6\frac{14}{15}$

Add the fractional parts.

$5\frac{4}{9} = 5\frac{20}{45}$ The LCM of 9 and 15 is 45.
$+6\frac{14}{15} = 6\frac{42}{45}$
$\phantom{+6\frac{14}{15} = } \frac{62}{45}$

Add the whole numbers.

$5\frac{4}{9} = 5\frac{20}{45}$
$+6\frac{14}{15} = 6\frac{42}{45}$
$11\frac{62}{45} = 11 + 1\frac{17}{45} = 12\frac{17}{45}$

Example 9   Add: $5 + \frac{3}{8}$

Solution $5 + \frac{3}{8} = 5\frac{3}{8}$

Example 10   Add: $7 + \frac{6}{11}$

Your solution

Example 11   Add: $17 + 3\frac{3}{8}$

Solution 17
$+ 3\frac{3}{8}$
$20\frac{3}{8}$

Example 12   Add: $29 + 17\frac{5}{12}$

Your solution

Example 13   Add: $5\frac{2}{3} + 11\frac{5}{6} + 12\frac{7}{9}$

Solution $5\frac{2}{3} = 5\frac{12}{18}$ LCM = 18.
$11\frac{5}{6} = 11\frac{15}{18}$
$+12\frac{7}{9} = 12\frac{14}{18}$
$\phantom{+12\frac{7}{9} = } 28\frac{41}{18} = 30\frac{5}{18}$

Example 14   Add: $7\frac{4}{5} + 6\frac{7}{10} + 13\frac{11}{15}$

Your solution

Example 15   Add: $11\frac{5}{8} + 7\frac{5}{9} + 8\frac{7}{15}$

Solution $11\frac{5}{8} = 11\frac{225}{360}$ LCM = 360.
$7\frac{5}{9} = 7\frac{200}{360}$
$+ 8\frac{7}{15} = 8\frac{168}{360}$
$\phantom{+18\frac{7}{15} = } 26\frac{593}{360} = 27\frac{233}{360}$

Example 16   Add: $9\frac{3}{8} + 17\frac{7}{12} + 10\frac{14}{15}$

Your solution

Solutions on p. A12

Objective D — To solve application problems

Example 17

A rain gauge collected $2\frac{1}{3}$ inches of rain in October, $5\frac{1}{2}$ inches in November, and $3\frac{3}{8}$ inches in December. Find the total rainfall for the 3 months.

Strategy

To find the total rainfall for the 3 months, add the 3 amounts of rainfall ($2\frac{1}{3}$, $5\frac{1}{2}$, and $3\frac{3}{8}$).

Solution

$$2\frac{1}{3} = 2\frac{8}{24}$$
$$5\frac{1}{2} = 5\frac{12}{24}$$
$$+3\frac{3}{8} = 3\frac{9}{24}$$
$$10\frac{29}{24} = 11\frac{5}{24}$$

The total rainfall for the three months was $11\frac{5}{24}$ inches.

Example 18

You spend $4\frac{1}{2}$ hours in class, $3\frac{3}{4}$ hours studying, and $1\frac{1}{3}$ hours eating each day. Find the total time spent each day on these three activities.

Your strategy

Your solution

Example 19

A student worked 4 hours, $2\frac{1}{3}$ hours, and $5\frac{2}{3}$ hours this week on a part-time job. The student is paid $6 an hour. How much did the student earn this week?

Strategy

To find how much the student earned:
- Find the total number of hours worked.
- Multiply the total number of hours worked by the hourly wage ($6).

Solution

$$\begin{array}{r} 4 \\ 2\frac{1}{3} \\ +5\frac{2}{3} \\ \hline 11\frac{3}{3} = 12 \text{ hours worked} \end{array} \qquad \begin{array}{r} 12 \\ \times \$6 \\ \hline \$72 \end{array}$$

The student earned $72 this week.

Example 20

A carpenter worked $1\frac{2}{3}$ hours of overtime on Monday, $3\frac{1}{3}$ hours of overtime on Tuesday, and 2 hours of overtime on Wednesday. At $24 an hour, find the carpenter's overtime pay for the 3 days.

Your strategy

Your solution

Solutions on p. A13

2.4 EXERCISES

▶ **Objective A**

Add.

1. $\frac{2}{7} + \frac{1}{7}$
2. $\frac{3}{11} + \frac{5}{11}$
3. $\frac{1}{2} + \frac{1}{2}$
4. $\frac{1}{3} + \frac{2}{3}$

5. $\frac{8}{11} + \frac{7}{11}$
6. $\frac{9}{13} + \frac{7}{13}$
7. $\frac{8}{5} + \frac{9}{5}$
8. $\frac{5}{3} + \frac{7}{3}$

9. $\frac{3}{5} + \frac{8}{5} + \frac{3}{5}$
10. $\frac{3}{8} + \frac{5}{8} + \frac{7}{8}$
11. $\frac{3}{4} + \frac{1}{4} + \frac{5}{4}$
12. $\frac{2}{7} + \frac{4}{7} + \frac{5}{7}$

13. $\frac{3}{8} + \frac{7}{8} + \frac{1}{8}$
14. $\frac{5}{12} + \frac{7}{12} + \frac{1}{12}$
15. $\frac{4}{15} + \frac{7}{15} + \frac{11}{15}$
16. $\frac{3}{4} + \frac{3}{4} + \frac{1}{4}$

17. $\frac{3}{16} + \frac{5}{16} + \frac{7}{16}$
18. $\frac{5}{18} + \frac{11}{18} + \frac{17}{18}$
19. $\frac{3}{11} + \frac{5}{11} + \frac{7}{11}$
20. $\frac{5}{7} + \frac{4}{7} + \frac{5}{7}$

▶ **Objective B**

Add.

21. $\frac{1}{2} + \frac{2}{3}$
22. $\frac{2}{3} + \frac{1}{4}$
23. $\frac{3}{14} + \frac{5}{7}$
24. $\frac{3}{5} + \frac{7}{10}$
25. $\frac{5}{9} + \frac{7}{15}$

26. $\frac{8}{15} + \frac{7}{20}$
27. $\frac{1}{6} + \frac{7}{9}$
28. $\frac{3}{8} + \frac{9}{14}$
29. $\frac{5}{12} + \frac{5}{16}$
30. $\frac{13}{40} + \frac{12}{25}$

31. $\frac{3}{20} + \frac{7}{30}$
32. $\frac{5}{12} + \frac{7}{30}$
33. $\frac{2}{3} + \frac{6}{19}$
34. $\frac{1}{2} + \frac{3}{29}$
35. $\frac{3}{7} + \frac{4}{21}$

36. $\frac{5}{14} + \frac{35}{84}$
37. $\frac{3}{14} + \frac{6}{49}$
38. $\frac{3}{10} + \frac{7}{45}$
39. $\frac{2}{7} + \frac{3}{8}$
40. $\frac{5}{8} + \frac{4}{9}$

Add.

41. $\frac{1}{3} + \frac{5}{6} + \frac{7}{9}$

42. $\frac{2}{3} + \frac{5}{6} + \frac{7}{12}$

43. $\frac{5}{6} + \frac{1}{12} + \frac{5}{16}$

44. $\frac{2}{9} + \frac{7}{15} + \frac{4}{21}$

45. $\frac{2}{3} + \frac{1}{5} + \frac{7}{12}$

46. $\frac{3}{4} + \frac{4}{5} + \frac{7}{12}$

47. $\frac{1}{4} + \frac{4}{5} + \frac{5}{9}$

48. $\frac{2}{3} + \frac{3}{5} + \frac{7}{8}$

49. $\frac{5}{16} + \frac{11}{18} + \frac{17}{24}$

50. $\frac{3}{10} + \frac{14}{15} + \frac{9}{25}$

51. $\frac{2}{3} + \frac{5}{8} + \frac{7}{9}$

52. $\frac{1}{3} + \frac{2}{9} + \frac{7}{8}$

53. $\frac{5}{6} + \frac{1}{2} + \frac{3}{11}$

54. $\frac{9}{11} + \frac{1}{2} + \frac{1}{6}$

55. $\frac{3}{4} + \frac{5}{6} + \frac{7}{8}$

▶ **Objective C**

Add.

56. $1\frac{1}{2}$
    $+2\frac{1}{6}$

57. $2\frac{2}{5}$
    $+3\frac{3}{10}$

58. $4\frac{1}{2}$
    $+5\frac{7}{12}$

59. $3\frac{3}{8}$
    $+2\frac{5}{16}$

60. $2\frac{7}{9}$
    $+3\frac{5}{12}$

61. $4\frac{7}{15}$
    $+3\frac{11}{12}$

62. 4
    $+5\frac{2}{7}$

63. $6\frac{8}{9}$
    $+12$

64. $3\frac{5}{8}$
    $+2\frac{11}{20}$

65. $4\frac{5}{12}$
    $+6\frac{11}{18}$

66. $10\frac{2}{7}$
    $+7\frac{21}{35}$

67. $16\frac{2}{3}$
    $+8\frac{1}{4}$

68. $16\frac{5}{8} + 3\frac{7}{20}$

69. $2\frac{3}{8} + 4\frac{9}{20}$

70. $2\frac{7}{10} + 7\frac{11}{15}$

71. $3\frac{3}{8} + 8\frac{5}{12}$

72. $7\frac{5}{12} + 2\frac{9}{16}$

73. $9\frac{1}{2} + 3\frac{3}{11}$

74. $6\frac{1}{3} + 2\frac{3}{13}$

75. $8\frac{21}{40} + 6\frac{21}{32}$

76. $8\frac{29}{30} + 7\frac{11}{40}$

77. $17\frac{5}{16} + 3\frac{11}{24}$

78. $17\frac{3}{8} + 7\frac{7}{20}$

79. $14\frac{7}{12} + 29\frac{13}{21}$

Add.

80. $5\frac{7}{8} + 27\frac{5}{12}$

81. $7\frac{5}{6} + 3\frac{5}{9}$

82. $7\frac{5}{9} + 2\frac{7}{12}$

83. $3\frac{1}{2} + 2\frac{3}{4} + 1\frac{5}{6}$

84. $2\frac{1}{2} + 3\frac{2}{3} + 4\frac{1}{4}$

85. $3\frac{1}{3} + 7\frac{1}{5} + 2\frac{1}{7}$

86. $3\frac{1}{2} + 3\frac{1}{5} + 8\frac{1}{9}$

87. $6\frac{5}{9} + 6\frac{5}{12} + 2\frac{5}{18}$

88. $2\frac{3}{8} + 4\frac{7}{12} + 3\frac{5}{16}$

89. $2\frac{1}{8} + 4\frac{2}{9} + 5\frac{17}{18}$

90. $6\frac{5}{6} + 17\frac{2}{9} + 18\frac{5}{27}$

91. $4\frac{7}{20} + \frac{17}{80} + 25\frac{23}{60}$

92. $7\frac{5}{12} + 1\frac{11}{24} + 9\frac{21}{36}$

93. $3\frac{5}{6} + 7\frac{7}{8} + 3\frac{1}{12}$

94. $7\frac{2}{5} + 3\frac{7}{10} + 5\frac{11}{15}$

▶ Objective D   **Application Problems**

95. A family of four find that $\frac{1}{3}$ of their income is spent on housing, $\frac{1}{8}$ is spent on transportation, and $\frac{1}{4}$ is spent on food. Find the total fractional amount of their income that is spent on these three items.

96. A table top that is $1\frac{1}{8}$ inch thick is covered by a $\frac{3}{16}$-inch veneer. Find the total thickness of the table top after the veneer is applied.

97. At the beginning of the year, the stock in a cellular phone company was selling for $\$26\frac{5}{8}$ per share. The price of the stock gained $\$13\frac{3}{4}$ per share during a 6-month period. Find the price of the stock at the end of 6 months.

98. At the beginning of the week, an airlines stock was selling at $\$158\frac{3}{8}$ per share. During the week, the stock gained $\$28\frac{3}{4}$ per share. Find the price of the stock at the end of the week.

99. Find the length of the shaft.

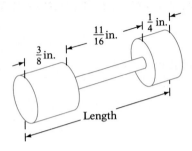

100. Find the length of the shaft.

101. You are working a part-time job that pays $7 an hour. You worked 5, $3\frac{3}{4}$, $2\frac{1}{3}$, $1\frac{1}{4}$, and $7\frac{2}{3}$ hours during the last 5 days.
  a. Find the total number of hours worked during the last 5 days.
  b. Find your total wages for the 5 days.

102. A nurse worked $2\frac{2}{3}$ hours of overtime on Monday, $1\frac{1}{4}$ hours on Wednesday, $1\frac{1}{3}$ hours on Friday, and $6\frac{3}{4}$ hours on Saturday.
  a. Find the total number of overtime hours worked during the week.
  b. At an overtime hourly wage of $22 an hour, how much overtime pay does the nurse receive?

103. A Southern California mountain had $5\frac{3}{4}$ inches of snow in December, $15\frac{1}{2}$ inches in January, and $9\frac{5}{8}$ inches in February. Find the total snowfall for the 3 months.

104. A utility stock was sold for $\$26\frac{3}{16}$. Monthly gains for 3 months were $\$1\frac{1}{2}$, $\$\frac{5}{16}$, and $\$2\frac{5}{8}$ per share. Find the value of the stock at the end of the 3 months.

Section 2.5 Subtraction of Fractions and Mixed Numbers

Objective A To subtract fractions with the same denominator

Fractions with the same denominator are subtracted by subtracting the numerators and placing the difference over the common denominator. After subtracting, write the fraction in simplest form.

Subtract: $\frac{5}{7} - \frac{3}{7}$

$$\begin{array}{r}\frac{5}{7}\\-\frac{3}{7}\\\hline\frac{2}{7}\end{array}$$

Subtract the numerators and place the difference over the common denominator.

$$\frac{5}{7} - \frac{3}{7} = \frac{5-3}{7} = \frac{2}{7}$$

Example 1 Subtract: $\frac{17}{30} - \frac{11}{30}$

Solution
$$\begin{array}{r}\frac{17}{30}\\-\frac{11}{30}\\\hline\frac{6}{30} = \frac{1}{5}\end{array}$$

Example 2 Subtract: $\frac{16}{27} - \frac{7}{27}$

Your solution

Solution on p. A13

Objective B To subtract fractions with unlike denominators

To subtract fractions with unlike denominators, first rewrite the fractions as equivalent fractions with a common denominator. As with adding fractions, the common denominator is the LCM of the denominators of the fractions.

Subtract: $\frac{5}{6} - \frac{1}{4}$

The common denominator is the LCM of 6 and 4. LCM = 12.

Build equivalent fractions using the LCM.

$$\frac{5}{6} = \frac{10}{12}$$
$$-\frac{1}{4} = \frac{3}{12}$$

Subtract the fractions.

$$\frac{5}{6} = \frac{10}{12}$$
$$-\frac{1}{4} = \frac{3}{12}$$
$$\overline{\phantom{-\frac{1}{4}=}\frac{7}{12}}$$

Example 3 Subtract: $\frac{2}{3} - \frac{1}{6}$

Solution
$$\frac{2}{3} = \frac{4}{6} \quad \text{LCM} = 6$$
$$-\frac{1}{6} = \frac{1}{6}$$
$$\overline{\frac{3}{6} = \frac{1}{2}}$$

Example 4 Subtract: $\frac{5}{6} - \frac{3}{8}$

Your solution

Example 5 Subtract: $\frac{11}{16} - \frac{5}{12}$

Solution
$$\frac{11}{16} = \frac{33}{48} \quad \text{LCM} = 48$$
$$-\frac{5}{12} = \frac{20}{48}$$
$$\overline{\frac{13}{48}}$$

Example 6 Subtract: $\frac{13}{18} - \frac{7}{24}$

Your solution

Solutions on p. A13

Objective C To subtract whole numbers, mixed numbers, and fractions

To subtract mixed numbers without borrowing, subtract the fractional parts and then subtract the whole numbers.

Subtract: $5\frac{5}{6} - 2\frac{3}{4}$

Subtract the fractional parts.

$$5\frac{5}{6} = 5\frac{10}{12} \quad \text{The LCM of 6 and 4 is 12.}$$
$$-2\frac{3}{4} = 2\frac{9}{12}$$
$$\overline{\phantom{-2\frac{3}{4} = 2}\frac{1}{12}}$$

Subtract the whole numbers.

$$5\frac{5}{6} = 5\frac{10}{12}$$
$$-2\frac{3}{4} = 2\frac{9}{12}$$
$$\overline{\phantom{-2\frac{3}{4} = 2}3\frac{1}{12}}$$

As with whole numbers, subtraction of mixed numbers sometimes involves borrowing.

Subtract: $5 - 2\frac{5}{8}$

Borrow 1 from 5.

$$5 = \overset{4}{\cancel{5}}1$$
$$-2\frac{5}{8} = 2\frac{5}{8}$$

Write 1 as a fraction so that the fractions have the same denominators.

$$5 = 4\frac{8}{8}$$
$$-2\frac{5}{8} = 2\frac{5}{8}$$

Subtract the mixed numbers.

$$5 = 4\frac{8}{8}$$
$$-2\frac{5}{8} = 2\frac{5}{8}$$
$$\overline{\phantom{-2\frac{5}{8} = }2\frac{3}{8}}$$

Subtract: $7\frac{1}{6} - 2\frac{5}{8}$

Write equivalent fractions using the LCM.	Borrow 1 from 7. Add the 1 to $\frac{4}{24}$. Write $1\frac{4}{24}$ as $\frac{28}{24}$.	Subtract the mixed numbers.
$7\frac{1}{6} = 7\frac{4}{24}$ $-2\frac{5}{8} = 2\frac{15}{24}$	$7\frac{1}{6} = 7\overset{6}{1}\frac{4}{24} = 6\frac{28}{24}$ $-2\frac{5}{8} = 2\frac{15}{24} = 2\frac{15}{24}$	$7\frac{1}{6} = 6\frac{28}{24}$ $-2\frac{5}{8} = 2\frac{15}{24}$ $\overline{4\frac{13}{24}}$

Example 7 Subtract: $15\frac{7}{8} - 12\frac{2}{3}$

Solution LCM = 24.

$15\frac{7}{8} = 15\frac{21}{24}$
$-12\frac{2}{3} = 12\frac{16}{24}$
$\phantom{-12\frac{2}{3} = 12}3\frac{5}{24}$

Example 8 Subtract: $17\frac{5}{9} - 11\frac{5}{12}$

Your solution

Example 9 Subtract: $9 - 4\frac{3}{11}$

Solution LCM = 11.

$9 \phantom{\frac{3}{11}} = 8\frac{11}{11}$
$-4\frac{3}{11} = 4\frac{3}{11}$
$\phantom{-4\frac{3}{11} =} 4\frac{8}{11}$

Example 10 Subtract: $8 - 2\frac{4}{13}$

Your solution

Example 11 Subtract: $11\frac{5}{12} - 2\frac{11}{16}$

Solution LCM = 48.

$11\frac{5}{12} = 11\frac{20}{48} = 10\frac{68}{48}$
$-2\frac{11}{16} = 2\frac{33}{48} = 2\frac{33}{48}$
$\phantom{-2\frac{11}{16} = 2\frac{33}{48} = }8\frac{35}{48}$

Example 12 Subtract: $21\frac{7}{9} - 7\frac{11}{12}$

Your solution

Example 13 Subtract: $72\frac{5}{42} - 49\frac{37}{63}$

Solution LCM = 126.

$72\frac{5}{42} = 72\frac{15}{126} = 71\frac{141}{126}$
$-49\frac{37}{63} = 49\frac{74}{126} = 49\frac{74}{126}$
$\phantom{-49\frac{37}{63} = 49\frac{74}{126} = }22\frac{67}{126}$

Example 14 Subtract: $81\frac{17}{80} - 17\frac{29}{80}$

Your solution

Solutions on p. A13

Objective D To solve application problems

Example 15

A $2\frac{2}{3}$-inch piece is cut from a $6\frac{5}{8}$-inch board. How much of the board is left?

Strategy

To find the length remaining, subtract the length of the piece cut from the total length of the board.

Solution

$$6\frac{5}{8} = 6\frac{15}{24} = 5\frac{39}{24}$$
$$-2\frac{2}{3} = 2\frac{16}{24} = 2\frac{16}{24}$$
$$\overline{\phantom{-2\frac{2}{3} = 2\frac{16}{24} = {}}3\frac{23}{24}}$$

$3\frac{23}{24}$ inches of the board are left.

Example 16

A flight from New York to Los Angeles takes $5\frac{1}{2}$ hours. After the plane is in the air for $2\frac{3}{4}$ hours, how much flight time remains?

Your strategy

Your solution

Example 17

Two painters are staining a house. In 1 day one painter stains $\frac{1}{3}$ of the house, and the other stains $\frac{1}{4}$ of the house. How much of the job remains to be done?

Strategy

To find how much of the job remains:
- Find the total amount of the house already stained $\left(\frac{1}{3} + \frac{1}{4}\right)$.
- Subtract the amount already stained from 1, which represents the complete job.

Solution

$$\begin{aligned}\frac{1}{3} &= \frac{4}{12} \\ +\frac{1}{4} &= \frac{3}{12} \\ \hline &\frac{7}{12}\end{aligned} \qquad \begin{aligned}1 &= \frac{12}{12} \\ -\frac{7}{12} &= \frac{7}{12} \\ \hline &\frac{5}{12}\end{aligned}$$

$\frac{5}{12}$ of the house remains to be stained.

Example 18

A patient is put on a diet to lose 24 pounds in 3 months. The patient loses $7\frac{1}{2}$ pounds the first month and $5\frac{3}{4}$ pounds the second month. How much weight must be lost the third month to achieve the goal?

Your strategy

Your solution

Solutions on pp. A13–A14

2.5 EXERCISES

Objective A

Subtract.

1. $\dfrac{9}{17} - \dfrac{7}{17}$
2. $\dfrac{11}{15} - \dfrac{3}{15}$
3. $\dfrac{11}{12} - \dfrac{7}{12}$
4. $\dfrac{13}{15} - \dfrac{4}{15}$
5. $\dfrac{9}{20} - \dfrac{7}{20}$
6. $\dfrac{27}{40} - \dfrac{13}{40}$

7. $\dfrac{48}{55} - \dfrac{13}{55}$
8. $\dfrac{42}{65} - \dfrac{17}{65}$
9. $\dfrac{11}{24} - \dfrac{5}{24}$
10. $\dfrac{23}{30} - \dfrac{13}{30}$
11. $\dfrac{17}{42} - \dfrac{5}{42}$
12. $\dfrac{29}{48} - \dfrac{13}{48}$

Objective B

Subtract.

13. $\dfrac{2}{3} - \dfrac{1}{6}$
14. $\dfrac{7}{8} - \dfrac{5}{16}$
15. $\dfrac{5}{8} - \dfrac{2}{7}$
16. $\dfrac{5}{6} - \dfrac{3}{7}$
17. $\dfrac{5}{7} - \dfrac{3}{14}$
18. $\dfrac{7}{10} - \dfrac{3}{5}$

19. $\dfrac{5}{9} - \dfrac{7}{15}$
20. $\dfrac{8}{15} - \dfrac{7}{20}$
21. $\dfrac{7}{9} - \dfrac{1}{6}$
22. $\dfrac{9}{14} - \dfrac{3}{8}$
23. $\dfrac{5}{12} - \dfrac{5}{16}$
24. $\dfrac{12}{25} - \dfrac{13}{40}$

25. $\dfrac{7}{30} - \dfrac{3}{20}$
26. $\dfrac{5}{12} - \dfrac{7}{30}$
27. $\dfrac{2}{3} - \dfrac{6}{19}$
28. $\dfrac{1}{2} - \dfrac{3}{29}$
29. $\dfrac{5}{9} - \dfrac{1}{12}$
30. $\dfrac{11}{16} - \dfrac{5}{12}$

Subtract.

31. $\dfrac{46}{51}$
    $-\dfrac{3}{17}$

32. $\dfrac{9}{16}$
    $-\dfrac{17}{32}$

33. $\dfrac{21}{35}$
    $-\dfrac{5}{14}$

34. $\dfrac{19}{40}$
    $-\dfrac{3}{16}$

35. $\dfrac{29}{60}$
    $-\dfrac{3}{40}$

36. $\dfrac{53}{70}$
    $-\dfrac{13}{35}$

37. $\dfrac{2}{9}$
    $-\dfrac{1}{42}$

38. $\dfrac{11}{15}$
    $-\dfrac{5}{27}$

39. $\dfrac{7}{16}$
    $-\dfrac{5}{24}$

40. $\dfrac{11}{18}$
    $-\dfrac{7}{24}$

41. $\dfrac{7}{8}$
    $-\dfrac{5}{9}$

42. $\dfrac{5}{6}$
    $-\dfrac{3}{5}$

43. $\dfrac{21}{29}$
    $-\dfrac{3}{8}$

44. $\dfrac{17}{31}$
    $-\dfrac{4}{9}$

▶ **Objective C**

Subtract.

45. $5\dfrac{7}{12}$
    $-2\dfrac{5}{12}$

46. $16\dfrac{11}{15}$
    $-11\dfrac{8}{15}$

47. $72\dfrac{21}{23}$
    $-16\dfrac{17}{23}$

48. $19\dfrac{16}{17}$
    $-9\dfrac{7}{17}$

49. $6\dfrac{1}{3}$
    -2

50. $5\dfrac{7}{8}$
    -1

51. 10
    $-6\dfrac{1}{3}$

52. 3
    $-2\dfrac{5}{21}$

53. $6\dfrac{2}{5}$
    $-4\dfrac{4}{5}$

54. $16\dfrac{3}{8}$
    $-10\dfrac{7}{8}$

55. $25\dfrac{4}{9}$
    $-16\dfrac{7}{9}$

56. $8\dfrac{3}{7}$
    $-2\dfrac{6}{7}$

57. $16\dfrac{2}{5}$
    $-8\dfrac{4}{9}$

58. $23\dfrac{7}{8}$
    $-16\dfrac{2}{3}$

59. 6
    $-4\dfrac{3}{5}$

Subtract.

60. $12 - 2\dfrac{7}{9}$

61. $16\dfrac{3}{20} - 1\dfrac{7}{20}$

62. $16\dfrac{3}{10} - 7\dfrac{9}{10}$

63. $25\dfrac{5}{12} - 4\dfrac{17}{24}$

64. $16\dfrac{5}{8} - 13\dfrac{11}{12}$

65. $6\dfrac{1}{8} - 3$

66. $32\dfrac{4}{15} - 17$

67. $32 - 6\dfrac{7}{15}$

68. $47 - 9\dfrac{7}{20}$

69. $14\dfrac{3}{5} - 7\dfrac{8}{9}$

70. $17\dfrac{3}{4} - 8\dfrac{10}{11}$

71. $3\dfrac{9}{32} - 1\dfrac{11}{40}$

72. $7\dfrac{11}{30} - 5\dfrac{13}{36}$

73. $9\dfrac{13}{40} - 7\dfrac{12}{25}$

74. $13\dfrac{27}{50} - 5\dfrac{39}{40}$

75. $65\dfrac{8}{35} - 16\dfrac{11}{14}$

76. $82\dfrac{4}{33} - 16\dfrac{5}{22}$

77. $101\dfrac{2}{9} - 16$

78. $77\dfrac{5}{18} - 61$

79. $17 - 7\dfrac{8}{13}$

80. $7 - 3\dfrac{22}{45}$

81. $25\dfrac{7}{15} - 14\dfrac{19}{36}$

82. $14\dfrac{17}{24} - 10\dfrac{31}{36}$

83. $61\dfrac{5}{21} - 28\dfrac{5}{9}$

84. $26\dfrac{7}{20} - 13\dfrac{11}{14}$

85. $137\dfrac{3}{5} - 69\dfrac{7}{12}$

86. $267\dfrac{4}{7} - 129\dfrac{2}{9}$

87. $13\dfrac{7}{25} - 6\dfrac{9}{20}$

88. $68\dfrac{5}{18} - 13\dfrac{17}{24}$

Objective D Application Problems

89. Molding $2\frac{2}{3}$ feet long is cut from an 8-foot piece of molding. What is the length of the remaining piece of molding?

90. The inseam of a pair of jeans is 30 inches long. After being washed the inseam of the jeans shrinks $\frac{5}{8}$ inch. Find the length of the pants after they are washed.

91. A plane trip from San Francisco to Washington takes $5\frac{1}{4}$ hours. After the plane is in the air $2\frac{1}{2}$ hours, how much time remains before landing?

92. A clerk can ride a bicycle to work in $\frac{3}{8}$ of an hour. The clerk can walk the distance in $1\frac{1}{4}$ hours. How much longer does it take to walk rather than ride to work?

93. A metal platform in a bridge is $280\frac{1}{2}$ feet during the summer. On a cold winter day the metal platform shrinks to $280\frac{3}{8}$ feet. How much did the bridge shrink due to the cold weather?

94. A housepainter painted $\frac{2}{5}$ of a house during 1 day. How much of the paint job remains to be done?

95. Find the missing dimension.

96. Find the missing dimension.

97. A 3-day $27\frac{1}{2}$-mile backpack hike is planned. The hikers plan to travel $7\frac{3}{8}$ miles the first day and $10\frac{1}{3}$ miles the second day.
    a. How many miles do the hikers plan to travel the first 2 days?
    b. How many miles are left to travel on the third day?

98. A 12-mile walkathon for charity has three checkpoints. The first checkpoint is $3\frac{3}{8}$ miles from the starting point. The second checkpoint is $4\frac{1}{3}$ miles from the first checkpoint.
    a. How many miles is it from the starting point to the second checkpoint?
    b. How many miles is it from the second checkpoint to the finish line?

99. A patient with high blood pressure is put on a diet to lose 38 pounds in 3 months. The patient loses $12\frac{1}{2}$ pounds the first month and $16\frac{3}{8}$ pounds the second month. How much weight must be lost the third month for the goal to be achieved?

100. To reach a desired weight of 172 pounds, a wrestler needs to lose $12\frac{3}{4}$ pounds. If $5\frac{1}{4}$ are lost the first week and $4\frac{1}{4}$ pounds the second week, how many more pounds must be lost for the desired weight to be reached?

Section 2.6 / Multiplication of Fractions and Mixed Numbers 91

SECTION 2.6 Multiplication of Fractions and Mixed Numbers

Objective A To multiply fractions

The product of two fractions is the product of the numerators over the product of the denominators.

Multiply: $\frac{2}{3} \times \frac{4}{5}$

Multiply the numerators.
Multiply the denominators.

$$\frac{2}{3} \times \frac{4}{5} = \frac{2 \cdot 4}{3 \cdot 5} = \frac{8}{15}$$

The product $\frac{2}{3} \times \frac{4}{5}$ can be read "$\frac{2}{3}$ times $\frac{4}{5}$" or "$\frac{2}{3}$ of $\frac{4}{5}$."

Reading the times sign as "of" is useful in application problems involving fractions and in diagramming the product of fractions.

$\frac{4}{5}$ of the bar is shaded.

Shade $\frac{2}{3}$ of the $\frac{4}{5}$ already shaded.

$\frac{8}{15}$ of the bar is then shaded.

$\frac{2}{3}$ of $\frac{4}{5} = \frac{2}{3} \times \frac{4}{5} = \frac{8}{15}$

After multiplying two fractions, write the product in simplest form.

Multiply: $\frac{3}{4} \times \frac{14}{15}$

Multiply the fractions.	Write the prime factorization of each number.	Eliminate the common factors.	Multiply the numbers remaining in the numerator. Multiply the numbers remaining in the denominator.

$$\frac{3}{4} \times \frac{14}{15} = \frac{3 \cdot 14}{4 \cdot 15} \quad = \frac{3 \cdot 2 \cdot 7}{2 \cdot 2 \cdot 3 \cdot 5} \quad = \frac{\overset{1}{\cancel{3}} \cdot \overset{1}{\cancel{2}} \cdot 7}{\underset{1}{\cancel{2}} \cdot 2 \cdot \underset{1}{\cancel{3}} \cdot 5} \quad \frac{7}{10}$$

Example 1

Multiply: $\frac{9}{10} \times \frac{2}{3}$

Solution

$$\frac{9}{10} \times \frac{2}{3} = \frac{9 \cdot 2}{10 \cdot 3} = \frac{\overset{1}{\cancel{3}} \cdot 3 \cdot \overset{1}{\cancel{2}}}{\underset{1}{\cancel{2}} \cdot 5 \cdot \underset{1}{\cancel{3}}} = \frac{3}{5}$$

Example 2

Multiply: $\frac{12}{25} \times \frac{5}{16}$

Your solution

Solution on p. A14

Example 3

Multiply: $\frac{4}{15} \times \frac{5}{28}$

Solution

$$\frac{4}{15} \times \frac{5}{28} = \frac{4 \cdot 5}{15 \cdot 28} = \frac{\overset{1}{\cancel{2}} \cdot \overset{1}{\cancel{2}} \cdot \overset{1}{\cancel{5}}}{3 \cdot \underset{1}{\cancel{5}} \cdot \underset{1}{\cancel{2}} \cdot \underset{1}{\cancel{2}} \cdot 7} = \frac{1}{21}$$

Example 4

Multiply: $\frac{4}{21} \times \frac{7}{44}$

Your solution

Example 5

Multiply: $\frac{9}{20} \times \frac{33}{35}$

Solution

$$\frac{9}{20} \times \frac{33}{35} = \frac{9 \cdot 33}{20 \cdot 35} = \frac{3 \cdot 3 \cdot 3 \cdot 11}{2 \cdot 2 \cdot 5 \cdot 5 \cdot 7} = \frac{297}{700}$$

Example 6

Multiply: $\frac{2}{21} \times \frac{10}{33}$

Your solution

Example 7

Multiply: $\frac{14}{9} \times \frac{12}{7}$

Solution

$$\frac{14}{9} \times \frac{12}{7} = \frac{14 \cdot 12}{9 \cdot 7} = \frac{2 \cdot \overset{1}{\cancel{7}} \cdot 2 \cdot 2 \cdot \overset{1}{\cancel{3}}}{3 \cdot \underset{1}{\cancel{3}} \cdot \underset{1}{\cancel{7}}} = \frac{8}{3} = 2\frac{2}{3}$$

Example 8

Multiply: $\frac{16}{5} \times \frac{15}{24}$

Your solution

Solution on p. A14

Objective B   To multiply whole numbers, mixed numbers, and fractions

To multiply a whole number by a fraction or mixed number, first write the whole number as a fraction with a denominator of 1.

Multiply: $4 \times \frac{3}{7}$

Write 4 with a denominator of 1; then multiply the fractions.

$$4 \times \frac{3}{7} = \frac{4}{1} \times \frac{3}{7} = \frac{4 \cdot 3}{1 \cdot 7} = \frac{2 \cdot 2 \cdot 3}{7} = \frac{12}{7} = 1\frac{5}{7}$$

When one or more of the factors in a product is a mixed number, write the mixed number as an improper fraction before multiplying.

Multiply: $2\frac{1}{3} \times \frac{3}{14}$

Write $2\frac{1}{3}$ as an improper fraction; then multiply the fractions.

$$2\frac{1}{3} \times \frac{3}{14} = \frac{7}{3} \times \frac{3}{14} = \frac{7 \cdot 3}{3 \cdot 14} = \frac{\overset{1}{\cancel{7}} \cdot \overset{1}{\cancel{3}}}{\underset{1}{\cancel{3}} \cdot 2 \cdot \underset{1}{\cancel{7}}} = \frac{1}{2}$$

Section 2.6 / Multiplication of Fractions and Mixed Numbers

Multiply: $5\frac{3}{5} \times 4\frac{2}{3}$

Write the mixed numbers as improper fractions; then multiply the fractions.

$$5\frac{3}{5} \times 4\frac{2}{3} = \frac{28}{5} \times \frac{14}{3} = \frac{28 \cdot 14}{5 \cdot 3} = \frac{2 \cdot 2 \cdot 7 \cdot 2 \cdot 7}{5 \cdot 3} = \frac{392}{15} = 26\frac{2}{15}$$

Example 9

Multiply: $4\frac{5}{6} \times \frac{12}{13}$

Solution

$$4\frac{5}{6} \times \frac{12}{13} = \frac{29}{6} \times \frac{12}{13} = \frac{29 \cdot 12}{6 \cdot 13}$$

$$= \frac{29 \cdot \cancel{2} \cdot 2 \cdot \cancel{3}}{\cancel{2} \cdot \cancel{3} \cdot 13} = \frac{58}{13} = 4\frac{6}{13}$$

Example 10

Multiply: $5\frac{2}{5} \times \frac{5}{9}$

Your solution

Example 11

Multiply: $\frac{3}{8} \times 6$

Solution

$$\frac{3}{8} \times 6 = \frac{3}{8} \times \frac{6}{1} = \frac{3 \cdot 6}{8 \cdot 1} = \frac{3 \cdot \cancel{2} \cdot 3}{2 \cdot \cancel{2} \cdot 2} = \frac{9}{4} = 2\frac{1}{4}$$

Example 12

Multiply: $9 \times \frac{7}{12}$

Your solution

Example 13

Multiply: $5\frac{2}{3} \times 4\frac{1}{2}$

Solution

$$5\frac{2}{3} \times 4\frac{1}{2} = \frac{17}{3} \times \frac{9}{2} = \frac{17 \cdot 9}{3 \cdot 2}$$

$$= \frac{17 \cdot \cancel{3} \cdot 3}{\cancel{3} \cdot 2} = \frac{51}{2} = 25\frac{1}{2}$$

Example 14

Multiply: $3\frac{2}{5} \times 6\frac{1}{4}$

Your solution

Example 15

Multiply: $4\frac{2}{5} \times 7$

Solution

$$4\frac{2}{5} \times 7 = \frac{22}{5} \times \frac{7}{1} = \frac{22 \cdot 7}{5 \cdot 1}$$

$$= \frac{2 \cdot 11 \cdot 7}{5} = \frac{154}{5} = 30\frac{4}{5}$$

Example 16

Multiply: $3\frac{2}{7} \times 6$

Your solution

Solutions on p. A14

Objective C To solve application problems

Example 17

An electrician earns $150 for each day worked. What are the electrician's earnings for working $4\frac{1}{2}$ days?

Strategy

To find the electrician's total earnings, multiply the daily earnings ($150) by the number of days worked $\left(4\frac{1}{2}\right)$.

Solution $150 \times 4\frac{1}{2} = \frac{150}{1} \times \frac{9}{2}$
$= \frac{150 \cdot 9}{1 \cdot 2}$
$= 675$

The electrician's earnings are $675.

Example 18

A house increased in value $3\frac{1}{2}$ times during the last 10 years. The price of the house 10 years ago was $30,000. What is the value of the house today?

Your strategy

Your solution

Example 19

The value of a small office building and the land on which it is built is $90,000. The value of the land is $\frac{1}{4}$ the total value. What is the value of the building (in dollars)?

Strategy

To find the value of the building:
- Find the value of the land $\left(\frac{1}{4} \times \$90,000\right)$.
- Subtract the value of the land from the total value.

Solution $\$90,000 \times \frac{1}{4} = \frac{\$90,000}{4}$
$= \$22,500$ value of the land

$\$90,000$
$-22,500$
$\$67,500$

The value of the building is $67,500.

Example 20

A paint company buys a drying chamber and an air compressor for spray painting. The total cost of the two items is $60,000. The drying chamber's cost is $\frac{4}{5}$ of the total cost. What is the cost of the air compressor?

Your strategy

Your solution

Solutions on p. A15

2.6 EXERCISES

▶ **Objective A**

Multiply.

1. $\dfrac{2}{3} \times \dfrac{7}{8}$
2. $\dfrac{1}{2} \times \dfrac{2}{3}$
3. $\dfrac{5}{16} \times \dfrac{7}{15}$
4. $\dfrac{3}{8} \times \dfrac{6}{7}$
5. $\dfrac{1}{2} \times \dfrac{5}{6}$

6. $\dfrac{1}{6} \times \dfrac{1}{8}$
7. $\dfrac{2}{5} \times \dfrac{5}{6}$
8. $\dfrac{11}{12} \times \dfrac{6}{7}$
9. $\dfrac{11}{12} \times \dfrac{3}{5}$
10. $\dfrac{2}{5} \times \dfrac{4}{9}$

11. $\dfrac{1}{6} \times \dfrac{6}{7}$
12. $\dfrac{3}{5} \times \dfrac{10}{11}$
13. $\dfrac{1}{5} \times \dfrac{5}{8}$
14. $\dfrac{6}{7} \times \dfrac{14}{15}$
15. $\dfrac{4}{9} \times \dfrac{15}{16}$

16. $\dfrac{8}{9} \times \dfrac{27}{4}$
17. $\dfrac{3}{5} \times \dfrac{3}{10}$
18. $\dfrac{5}{6} \times \dfrac{1}{2}$
19. $\dfrac{3}{8} \times \dfrac{5}{12}$
20. $\dfrac{3}{10} \times \dfrac{6}{7}$

21. $\dfrac{16}{9} \times \dfrac{27}{8}$
22. $\dfrac{5}{8} \times \dfrac{16}{15}$
23. $\dfrac{3}{2} \times \dfrac{4}{9}$
24. $\dfrac{5}{3} \times \dfrac{3}{7}$
25. $\dfrac{5}{16} \times \dfrac{12}{11}$

26. $\dfrac{7}{8} \times \dfrac{3}{14}$
27. $\dfrac{2}{9} \times \dfrac{1}{5}$
28. $\dfrac{1}{10} \times \dfrac{3}{8}$
29. $\dfrac{5}{12} \times \dfrac{6}{7}$
30. $\dfrac{1}{3} \times \dfrac{9}{2}$

31. $\dfrac{15}{8} \times \dfrac{16}{3}$
32. $\dfrac{5}{6} \times \dfrac{4}{15}$
33. $\dfrac{1}{2} \times \dfrac{2}{15}$
34. $\dfrac{3}{8} \times \dfrac{5}{16}$
35. $\dfrac{2}{9} \times \dfrac{15}{17}$

36. $\dfrac{5}{7} \times \dfrac{14}{15}$
37. $\dfrac{3}{8} \times \dfrac{15}{41}$
38. $\dfrac{5}{12} \times \dfrac{42}{65}$
39. $\dfrac{16}{33} \times \dfrac{55}{72}$
40. $\dfrac{8}{3} \times \dfrac{21}{32}$

41. $\dfrac{12}{5} \times \dfrac{5}{3}$
42. $\dfrac{17}{9} \times \dfrac{81}{17}$
43. $\dfrac{16}{85} \times \dfrac{125}{84}$
44. $\dfrac{19}{64} \times \dfrac{48}{95}$

▶ **Objective B**

Multiply.

45. $4 \times \frac{3}{8}$

46. $14 \times \frac{5}{7}$

47. $\frac{2}{3} \times 6$

48. $\frac{5}{12} \times 40$

49. $\frac{1}{3} \times 1\frac{1}{3}$

50. $\frac{2}{5} \times 2\frac{1}{2}$

51. $1\frac{7}{8} \times \frac{4}{15}$

52. $2\frac{1}{5} \times \frac{5}{22}$

53. $55 \times \frac{3}{10}$

54. $\frac{5}{14} \times 49$

55. $4 \times 2\frac{1}{2}$

56. $9 \times 3\frac{1}{3}$

57. $2\frac{1}{7} \times 3$

58. $5\frac{1}{4} \times 8$

59. $3\frac{2}{3} \times 5$

60. $4\frac{2}{9} \times 3$

61. $\frac{1}{2} \times 3\frac{3}{7}$

62. $\frac{3}{8} \times 4\frac{4}{5}$

63. $6\frac{1}{8} \times \frac{4}{7}$

64. $5\frac{1}{3} \times \frac{5}{16}$

65. $5\frac{1}{8} \times 5$

66. $6\frac{1}{9} \times 2$

67. $\frac{3}{8} \times 4\frac{1}{2}$

68. $\frac{5}{7} \times 2\frac{1}{3}$

69. $6 \times 2\frac{2}{3}$

70. $6\frac{1}{8} \times 0$

71. $1\frac{1}{3} \times 2\frac{1}{4}$

72. $2\frac{5}{8} \times \frac{3}{23}$

73. $1\frac{1}{2} \times 5\frac{1}{2}$

74. $3\frac{1}{8} \times 2\frac{1}{4}$

75. $6\frac{1}{2} \times 4\frac{3}{13}$

76. $7\frac{3}{5} \times 4\frac{1}{6}$

77. $5\frac{1}{2} \times 1\frac{3}{11}$

78. $2\frac{2}{5} \times 1\frac{7}{12}$

79. $3\frac{1}{2} \times 7$

80. $5\frac{2}{3} \times 12$

81. $3\frac{1}{3} \times 6\frac{3}{5}$

82. $5\frac{3}{7} \times 3\frac{1}{2}$

83. $4\frac{3}{13} \times 7\frac{4}{5}$

84. $7\frac{5}{12} \times 3\frac{3}{7}$

Multiply.

85. $2\frac{5}{8} \times 3\frac{2}{5}$
86. $5\frac{3}{16} \times 5\frac{1}{3}$
87. $3\frac{1}{7} \times 2\frac{1}{8}$
88. $16\frac{5}{8} \times 1\frac{1}{16}$

89. $9\frac{5}{8} \times 7\frac{6}{7}$
90. $2\frac{1}{3} \times 3\frac{1}{3}$
91. $2 \times 4\frac{1}{8}$
92. $3\frac{2}{3} \times 5\frac{2}{11}$

93. $5 \times \frac{2}{3}$
94. $0 \times 5\frac{2}{3}$
95. $2\frac{2}{7} \times 3\frac{1}{2}$
96. $4\frac{1}{2} \times \frac{1}{3}$

97. $2\frac{2}{3} \times 3\frac{1}{4}$
98. $\frac{3}{4} \times 5\frac{1}{3}$
99. $5\frac{1}{5} \times 3\frac{3}{4}$
100. $4\frac{1}{7} \times 2\frac{1}{3}$

101. $\frac{5}{8} \times 5\frac{3}{5}$
102. $6\frac{1}{3} \times 1\frac{2}{19}$
103. $1\frac{1}{2} \times 1\frac{1}{3}$
104. $3\frac{5}{8} \times 0$

105. $2\frac{2}{5} \times 3\frac{1}{12}$
106. $2\frac{2}{3} \times \frac{3}{20}$
107. $5\frac{1}{5} \times 3\frac{1}{13}$
108. $3\frac{3}{4} \times 2\frac{3}{20}$

109. $10\frac{1}{4} \times 3\frac{1}{5}$
110. $12\frac{3}{5} \times 1\frac{3}{7}$
111. $5\frac{3}{7} \times 5\frac{1}{4}$
112. $6\frac{1}{2} \times 1\frac{3}{13}$

113. $8\frac{2}{3} \times 2\frac{1}{13}$
114. $10\frac{2}{5} \times 3\frac{5}{26}$
115. $3\frac{1}{5} \times 2\frac{2}{3}$
116. $15\frac{1}{3} \times 6\frac{1}{2}$

117. $5\frac{3}{8} \times 6\frac{2}{5}$
118. $3\frac{1}{2} \times 2\frac{5}{14}$
119. $21\frac{5}{8} \times 24$
120. $16 \times 3\frac{3}{4}$

121. $84 \times 3\frac{3}{14}$
122. $5\frac{1}{8} \times 41$
123. $11\frac{1}{3} \times 7\frac{3}{34}$
124. $8\frac{7}{9} \times 6\frac{6}{11}$

▶ **Objective C** *Application Problems*

125. A car can travel $23\frac{1}{2}$ miles on 1 gallon of gas. How far can the car travel on $\frac{1}{2}$ gallon of gas?

126. A person can walk $3\frac{1}{2}$ miles in 1 hour. How far can the person walk in $\frac{1}{3}$ hour?

127. A compact car travels 38 miles on each gallon of gasoline. How many miles can the car travel on $9\frac{1}{2}$ gallons of gasoline?

128. A board is $5\frac{3}{4}$ feet long. One third of the board is cut off. What is the length of the piece cut off?

129. A cook is using a recipe that calls for $3\frac{1}{2}$ cups of flour. The cook wants to double the recipe. How much flour should the cook use?

130. The F-1 engine in the first stage of the Saturn 5 rocket burns 214,000 gallons of propellant in 1 minute. The first stage burns $2\frac{1}{2}$ minutes before burnout. How much propellant is used before burnout?

131. The stock of a high technology company is selling for $\$68\frac{3}{8}$. Find the cost of 40 shares of the stock.

132. A soft drink company budgets $\frac{1}{10}$ of its income each month for advertising. In July the company had an income of $24,500. What is the amount budgeted for advertising in July?

133. A family budgets $\frac{2}{5}$ of its monthly income of $3200 per month for housing and utilities.
    a. What amount is budgeted for housing and utilities?
    b. What amount remains for purposes other than housing and utilities?

134. A student read $\frac{2}{3}$ of a book containing 432 pages.
    a. How many pages did the student read?
    b. How many pages remain to be read?

135. The parents of the school choir members are making robes for the choir. Each robe requires $2\frac{5}{8}$ yards of material at a cost of $8 per yard. Find the total cost of 24 choir robes.

136. A college spends $\frac{5}{8}$ of its monthly income on employee salaries. During one month the college had an income of $712,000. How much of the monthly income remained after the employee's salaries were paid?

SECTION 2.7 Division of Fractions and Mixed Numbers

Objective A To divide fractions

The **reciprocal** of a fraction is the fraction with the numerator and denominator interchanged. The process of interchanging the numerator and denominator of a fraction is called **inverting** the fraction.

The reciprocal of $\frac{2}{3}$ is $\frac{3}{2}$.

To find the reciprocal of a whole number, first write the whole number as a fraction with a denominator of 1; then find the reciprocal of that fraction.

The reciprocal of 5 is $\frac{1}{5}$. $\left(\text{Think } 5 = \frac{5}{1}\right)$

Reciprocals are used to rewrite division problems as related multiplication problems. Look at the following two problems:

$$8 \div 2 = 4 \qquad\qquad 8 \times \frac{1}{2} = 4$$

8 divided by 2 is 4.   8 times the reciprocal of 2 is 4.

"Divided by" means the same as "times the reciprocal of." Thus "÷ 2" can be replaced with "× $\frac{1}{2}$," and the answer will be the same. Fractions are divided by making this replacement.

Divide: $\frac{2}{3} \div \frac{3}{4}$ $\frac{2}{3} \div \frac{3}{4} = \frac{2}{3} \times \frac{4}{3} = \frac{2 \cdot 4}{3 \cdot 3} = \frac{2 \cdot 2 \cdot 2}{3 \cdot 3} = \frac{8}{9}$

Example 1   Divide: $\frac{5}{8} \div \frac{4}{9}$

Solution $\frac{5}{8} \div \frac{4}{9} = \frac{5}{8} \times \frac{9}{4} = \frac{5 \cdot 9}{8 \cdot 4}$
$= \frac{5 \cdot 3 \cdot 3}{2 \cdot 2 \cdot 2 \cdot 2 \cdot 2} = \frac{45}{32} = 1\frac{13}{32}$

Example 2   Divide: $\frac{3}{7} \div \frac{2}{3}$

Your solution

Example 3   Divide: $\frac{3}{5} \div \frac{12}{25}$

Solution $\frac{3}{5} \div \frac{12}{25} = \frac{3}{5} \times \frac{25}{12} = \frac{3 \cdot 25}{5 \cdot 12}$
$= \frac{\cancel{3} \cdot \cancel{5} \cdot 5}{\cancel{5} \cdot 2 \cdot 2 \cdot \cancel{3}} = \frac{5}{4} = 1\frac{1}{4}$

Example 4   Divide: $\frac{3}{4} \div \frac{9}{10}$

Your solution

Solutions on p. A15

Objective B To divide whole numbers, mixed numbers, and fractions

To divide a fraction and a whole number, first write the whole number as a fraction with a denominator of 1.

Divide: $\frac{3}{7} \div 5$

Write 5 with a denominator of 1; $\frac{3}{7} \div \boxed{5} = \frac{3}{7} \div \boxed{\frac{5}{1}} = \frac{3}{7} \times \frac{1}{5} = \frac{3 \cdot 1}{7 \cdot 5} = \frac{3}{35}$
then divide the fractions.

When one of the numbers in a quotient is a mixed number, write the mixed number as an improper fraction before dividing.

Divide: $4\frac{2}{3} \div \frac{8}{15}$

Write $4\frac{2}{3}$ as an improper fraction; then divide the fractions.

$$4\frac{2}{3} \div \frac{8}{15} = \frac{14}{3} \div \frac{8}{15} = \frac{14}{3} \times \frac{15}{8} = \frac{14 \cdot 15}{3 \cdot 8} = \frac{\cancel{2} \cdot 7 \cdot \cancel{3} \cdot 5}{\cancel{3} \cdot 2 \cdot \cancel{2} \cdot 2} = \frac{35}{4} = 8\frac{3}{4}$$

Divide: $1\frac{13}{15} \div 4\frac{4}{5}$

Write the mixed numbers as improper fractions. Then divide the fractions.

$$1\frac{13}{15} \div 4\frac{4}{5} = \frac{28}{15} \div \frac{24}{5} = \frac{28}{15} \times \frac{5}{24} = \frac{28 \cdot 5}{15 \cdot 24} = \frac{\cancel{2} \cdot \cancel{2} \cdot 7 \cdot \cancel{5}}{3 \cdot \cancel{5} \cdot \cancel{2} \cdot \cancel{2} \cdot 2 \cdot 3} = \frac{7}{18}$$

Example 5 Divide: $6 \div \frac{3}{4}$

Solution

$$6 \div \frac{3}{4} = \frac{6}{1} \div \frac{3}{4} = \frac{6}{1} \times \frac{4}{3}$$

$$= \frac{2 \cdot \cancel{3} \cdot 2 \cdot 2}{\cancel{3}} = \frac{8}{1} = 8$$

Example 6 Divide: $8 \div \frac{2}{3}$

Your solution

Example 7 Divide: $\frac{4}{9} \div 5$

Solution

$$\frac{4}{9} \div 5 = \frac{4}{9} \div \frac{5}{1} = \frac{4}{9} \times \frac{1}{5}$$

$$= \frac{4 \cdot 1}{9 \cdot 5} = \frac{2 \cdot 2}{3 \cdot 3 \cdot 5} = \frac{4}{45}$$

Example 8 Divide: $\frac{5}{7} \div 6$

Your solution

Example 9 Divide: $\frac{3}{8} \div 2\frac{1}{10}$

Solution

$$\frac{3}{8} \div 2\frac{1}{10} = \frac{3}{8} \div \frac{21}{10} = \frac{3}{8} \times \frac{10}{21}$$

$$= \frac{3 \cdot 10}{8 \cdot 21} = \frac{\cancel{3} \cdot \cancel{2} \cdot 5}{\cancel{2} \cdot 2 \cdot 2 \cdot \cancel{3} \cdot 7} = \frac{5}{28}$$

Example 10 Divide: $12\frac{3}{5} \div 7$

Your solution

Example 11 Divide: $2\frac{3}{4} \div 1\frac{5}{7}$

Solution

$$2\frac{3}{4} \div 1\frac{5}{7} = \frac{11}{4} \div \frac{12}{7} = \frac{11}{4} \times \frac{7}{12} = \frac{11 \cdot 7}{4 \cdot 12}$$

$$= \frac{11 \cdot 7}{2 \cdot 2 \cdot 2 \cdot 2 \cdot 3} = \frac{77}{48} = 1\frac{29}{48}$$

Example 12 Divide: $3\frac{2}{3} \div 2\frac{2}{5}$

Your solution

Solutions on p. A15

Example 13 Divide: $1\frac{13}{15} \div 4\frac{1}{5}$

Solution

$$1\frac{13}{15} \div 4\frac{1}{5} = \frac{28}{15} \div \frac{21}{5} = \frac{28}{15} \times \frac{5}{21} = \frac{28 \cdot 5}{15 \cdot 21}$$

$$= \frac{2 \cdot 2 \cdot \cancel{7} \cdot \cancel{5}}{3 \cdot \cancel{5} \cdot 3 \cdot \cancel{7}} = \frac{4}{9}$$

Example 14 Divide: $2\frac{5}{6} \div 8\frac{1}{2}$

Your solution

Example 15 Divide: $4\frac{3}{8} \div 7$

Solution

$$4\frac{3}{8} \div 7 = \frac{35}{8} \div \frac{7}{1} = \frac{35}{8} \times \frac{1}{7}$$

$$= \frac{35 \cdot 1}{8 \cdot 7} = \frac{5 \cdot \cancel{7}}{2 \cdot 2 \cdot 2 \cdot \cancel{7}} = \frac{5}{8}$$

Example 16 Divide: $6\frac{2}{5} \div 4$

Your solution

Solutions on p. A15

Objective C To solve application problems

Example 17

A car used $15\frac{1}{2}$ gallons of gasoline on a 310-mile trip. How many miles can this car travel on 1 gallon of gasoline?

Strategy

To find the number of miles, divide the number of miles traveled by the number of gallons of gasoline used.

Solution

$$310 \div 15\frac{1}{2} = \frac{310}{1} \div \frac{31}{2}$$

$$= \frac{310}{1} \times \frac{2}{31}$$

$$= \frac{310 \cdot 2}{1 \cdot 31}$$

$$= 20$$

The car travels 20 miles on 1 gallon of gasoline.

Example 18

An investor purchased a $2\frac{1}{2}$-ounce gold coin for $975. What was the price for 1 ounce?

Your strategy

Your solution

Solution on p. A15

Example 19

A 12-foot board is cut into pieces $2\frac{1}{4}$ feet long for use as bookshelves. What is the length of the remaining piece after as many shelves as possible are cut?

Strategy

To find the length of the remaining piece:
- Divide the total length by the length of each shelf $\left(12 \div 2\frac{1}{4}\right)$. This will give you the number of shelves cut, with a certain fraction of a shelf left over.
- Multiply the fraction left over by the length of a shelf to determine the actual length of the piece remaining.

Solution

$$12 \div 2\frac{1}{4} = \frac{12}{1} \div \frac{9}{4} = \frac{12}{1} \times \frac{4}{9}$$
$$= \frac{12 \cdot 4}{1 \cdot 9} = \frac{16}{3} = 5\frac{1}{3}$$

5 pieces $2\frac{1}{4}$ feet long

1 piece $\frac{1}{3}$ of $2\frac{1}{4}$ feet long

$$\frac{1}{3} \times 2\frac{1}{4} = \frac{1}{3} \times \frac{9}{4} = \frac{1 \cdot 9}{3 \cdot 4} = \frac{3}{4}$$

The length of the piece remaining is $\frac{3}{4}$ foot.

Example 20

A 16-foot board is cut into pieces $3\frac{1}{3}$ feet long for shelves for a bookcase. What is the length of the remaining piece after as many shelves as possible are cut?

Your strategy

Your solution

Solution on p. A15

2.7 EXERCISES

▶ **Objective A**

Divide.

1. $\dfrac{1}{3} \div \dfrac{2}{5}$
2. $\dfrac{3}{7} \div \dfrac{3}{2}$
3. $\dfrac{31}{40} \div \dfrac{11}{15}$
4. $\dfrac{3}{7} \div \dfrac{3}{7}$
5. $0 \div \dfrac{1}{2}$

6. $0 \div \dfrac{3}{4}$
7. $\dfrac{16}{33} \div \dfrac{4}{11}$
8. $\dfrac{5}{24} \div \dfrac{15}{36}$
9. $\dfrac{11}{15} \div \dfrac{1}{12}$
10. $\dfrac{2}{9} \div \dfrac{16}{19}$

11. $\dfrac{15}{16} \div \dfrac{16}{39}$
12. $\dfrac{2}{15} \div \dfrac{3}{5}$
13. $\dfrac{8}{9} \div \dfrac{4}{5}$
14. $\dfrac{11}{15} \div \dfrac{5}{22}$
15. $\dfrac{12}{13} \div \dfrac{4}{9}$

16. $\dfrac{1}{9} \div \dfrac{2}{3}$
17. $\dfrac{10}{21} \div \dfrac{5}{7}$
18. $\dfrac{2}{5} \div \dfrac{4}{7}$
19. $\dfrac{3}{8} \div \dfrac{5}{12}$
20. $\dfrac{5}{9} \div \dfrac{15}{32}$

21. $\dfrac{1}{2} \div \dfrac{1}{4}$
22. $\dfrac{1}{3} \div \dfrac{1}{9}$
23. $\dfrac{1}{5} \div \dfrac{1}{10}$
24. $\dfrac{4}{15} \div \dfrac{2}{5}$

25. $\dfrac{7}{15} \div \dfrac{14}{5}$
26. $\dfrac{5}{8} \div \dfrac{15}{2}$
27. $\dfrac{14}{3} \div \dfrac{7}{9}$
28. $\dfrac{7}{4} \div \dfrac{9}{2}$

29. $\dfrac{5}{9} \div \dfrac{25}{3}$
30. $\dfrac{5}{16} \div \dfrac{3}{8}$
31. $\dfrac{2}{3} \div \dfrac{1}{3}$
32. $\dfrac{4}{9} \div \dfrac{1}{9}$

33. $\dfrac{5}{7} \div \dfrac{2}{7}$
34. $\dfrac{5}{6} \div \dfrac{1}{9}$
35. $\dfrac{2}{3} \div \dfrac{2}{9}$
36. $\dfrac{5}{12} \div \dfrac{5}{6}$

37. $4 \div \dfrac{2}{3}$
38. $\dfrac{2}{3} \div 4$
39. $\dfrac{3}{2} \div 3$
40. $3 \div \dfrac{3}{2}$

Divide.

41. $\dfrac{15}{8} \div \dfrac{5}{16}$ **42.** $\dfrac{15}{8} \div \dfrac{5}{32}$ **43.** $16 \div \dfrac{24}{31}$ **44.** $\dfrac{8}{19} \div \dfrac{7}{38}$

▶ **Objective B**

Divide.

45. $\dfrac{5}{6} \div 25$ **46.** $22 \div \dfrac{3}{11}$ **47.** $6 \div 3\dfrac{1}{3}$ **48.** $5\dfrac{1}{2} \div 11$ **49.** $3\dfrac{1}{3} \div \dfrac{3}{8}$

50. $6\dfrac{1}{2} \div \dfrac{1}{2}$ **51.** $\dfrac{3}{8} \div 2\dfrac{1}{4}$ **52.** $\dfrac{5}{12} \div 4\dfrac{4}{5}$ **53.** $1\dfrac{1}{2} \div 1\dfrac{3}{8}$ **54.** $2\dfrac{1}{4} \div 1\dfrac{3}{8}$

55. $1\dfrac{3}{5} \div 2\dfrac{1}{10}$ **56.** $2\dfrac{5}{6} \div 1\dfrac{1}{9}$ **57.** $2\dfrac{1}{3} \div 3\dfrac{2}{3}$ **58.** $4\dfrac{1}{2} \div 2\dfrac{1}{6}$ **59.** $7\dfrac{1}{2} \div 2\dfrac{2}{3}$

60. $8\dfrac{1}{4} \div 2\dfrac{3}{4}$ **61.** $3\dfrac{5}{9} \div 32$ **62.** $4\dfrac{1}{5} \div 21$ **63.** $6\dfrac{8}{9} \div \dfrac{31}{36}$ **64.** $5\dfrac{3}{5} \div \dfrac{7}{10}$

65. $\dfrac{11}{12} \div 2\dfrac{1}{3}$ **66.** $\dfrac{7}{8} \div 3\dfrac{1}{4}$ **67.** $\dfrac{5}{16} \div 5\dfrac{3}{8}$ **68.** $\dfrac{9}{14} \div 3\dfrac{1}{7}$ **69.** $27 \div \dfrac{3}{8}$

70. $35 \div \dfrac{7}{24}$ **71.** $\dfrac{3}{8} \div 2\dfrac{3}{4}$ **72.** $\dfrac{11}{18} \div 2\dfrac{2}{9}$ **73.** $\dfrac{21}{40} \div 3\dfrac{3}{10}$ **74.** $\dfrac{6}{25} \div 4\dfrac{1}{5}$

75. $2\dfrac{1}{16} \div 2\dfrac{1}{2}$ **76.** $7\dfrac{3}{5} \div 1\dfrac{7}{12}$ **77.** $1\dfrac{2}{3} \div \dfrac{3}{8}$ **78.** $16 \div \dfrac{2}{3}$ **79.** $2\dfrac{1}{2} \div 3\dfrac{5}{14}$

80. $1\dfrac{5}{8} \div 4$ **81.** $13\dfrac{3}{8} \div \dfrac{1}{4}$ **82.** $16 \div 1\dfrac{1}{2}$ **83.** $9 \div \dfrac{7}{8}$ **84.** $15\dfrac{1}{3} \div 2\dfrac{2}{9}$

Divide.

85. $16\frac{5}{8} \div 1\frac{2}{3}$
86. $24\frac{4}{5} \div 2\frac{3}{5}$
87. $1\frac{1}{3} \div 5\frac{8}{9}$
88. $13\frac{2}{3} \div 0$

89. $15\frac{1}{3} \div 7\frac{1}{6}$
90. $2\frac{3}{49} \div 1\frac{11}{14}$
91. $0 \div 1\frac{1}{2}$
92. $3\frac{5}{12} \div 2\frac{1}{6}$

93. $4\frac{5}{16} \div 2\frac{3}{8}$
94. $17\frac{4}{5} \div 9\frac{3}{10}$
95. $6\frac{1}{3} \div 5$
96. $15 \div \frac{1}{8}$

97. $\frac{1}{12} \div 12$
98. $14\frac{3}{8} \div 6\frac{7}{12}$
99. $3\frac{5}{11} \div 3\frac{4}{5}$
100. $1\frac{25}{36} \div 2\frac{11}{24}$

101. $19\frac{1}{2} \div \frac{1}{2}$
102. $\frac{1}{4} \div 2\frac{5}{6}$
103. $5\frac{3}{8} \div 2\frac{7}{16}$
104. $7\frac{4}{13} \div 2\frac{3}{26}$

105. $82\frac{3}{4} \div 6\frac{1}{8}$
106. $21\frac{5}{14} \div 7\frac{3}{28}$
107. $19\frac{3}{7} \div 4\frac{5}{14}$
108. $16\frac{5}{21} \div 3\frac{9}{42}$

109. $82\frac{3}{5} \div 19\frac{1}{10}$
110. $45\frac{3}{5} \div 15$
111. $102 \div 1\frac{1}{2}$
112. $44\frac{5}{12} \div 32\frac{5}{6}$

113. $0 \div 3\frac{1}{2}$
114. $8\frac{2}{7} \div 1$
115. $6\frac{9}{16} \div 1\frac{3}{32}$
116. $5\frac{5}{14} \div 3\frac{4}{7}$

117. $8\frac{8}{9} \div 2\frac{13}{18}$
118. $10\frac{1}{5} \div 1\frac{7}{10}$
119. $7\frac{3}{8} \div 1\frac{27}{32}$
120. $4\frac{5}{11} \div 2\frac{1}{3}$

121. $15\frac{3}{8} \div 7\frac{3}{4}$
122. $7\frac{2}{9} \div 1\frac{17}{18}$
123. $3\frac{7}{8} \div 7\frac{3}{4}$
124. $9\frac{7}{12} \div \frac{5}{12}$

▶ **Objective C** *Application Problems*

125. Individual cereal boxes contain $\frac{3}{4}$ ounce of cereal. How many boxes can be filled with 600 ounces of cereal?

126. A box contains 20 ounces of cereal. How many $1\frac{1}{4}$-ounce portions can be served from this box?

127. A sales executive used $18\frac{2}{5}$ gallons of gas on a 460-mile trip. How many miles can the sales executive travel on 1 gallon of gas?

128. A commuter plane used $63\frac{3}{4}$ gallons of fuel on a 5-hour flight. How many gallons were used each hour?

129. A $\frac{5}{8}$-carat diamond was purchased for $1200. What would a similar diamond weighing 1 carat cost?

130. An investor bought $8\frac{1}{3}$ acres of land for $200,000. What was the cost of each acre?

131. A utility stock is offered for $$4\frac{3}{4}$ per share. How many shares can you buy for $285?

132. A new play opened, and 1200 people attended the opening at the concert hall. The hall was $\frac{2}{3}$ full. What is the capacity of the concert hall?

133. A contractor purchased $9\frac{3}{4}$ acres of land for a building project. One and one-half acres were set aside for a park.
    a. How many acres are available for building?
    b. How many $\frac{1}{4}$-acre parcels of land can be sold after the land for the park is removed?

134. A chef purchased a roast that weighed $10\frac{3}{4}$ pounds. After the fat was trimmed and the bone removed, the roast weighed $9\frac{1}{3}$ pounds.
    a. What was the total weight of the fat and bone?
    b. How many $\frac{1}{3}$-pound meat servings can the chef cut from the roast?

135. One-tenth of a shipment of $18\frac{1}{3}$ pounds of grapes were spoiled. How many $\frac{1}{2}$-pound packages of unspoiled grapes can be packaged from this shipment?

136. A 15-foot board is cut into pieces $3\frac{1}{2}$ feet long for a bookcase. What is the length of the piece remaining after as many shelves as possible have been cut?

SECTION 2.8 Order, Exponents, and the Order of Operations Agreement

Objective A To identify the order relation between two fractions

Recall that whole numbers can be graphed as points on the number line. Fractions can also be graphed as points on the number line.

The graph of $\frac{3}{4}$ on the number line

The number line can be used to determine the order relation between two fractions. A fraction that appears to the left of a given fraction is less than the given fraction. A fraction that appears to the right of a given fraction is greater than the given fraction.

$\frac{1}{8} < \frac{3}{8}$ $\frac{6}{8} > \frac{3}{8}$

To find the order relation between two fractions with the same denominator, compare numerators. The fraction that has the smaller numerator is the smaller fraction. When the denominators are different, begin by writing equivalent fractions with a common denominator; then compare numerators.

Find the order relation between $\frac{11}{18}$ and $\frac{5}{8}$.

The LCM of 18 and 8 is 72.

$\frac{11}{18} = \frac{44}{72}$ ← Smaller Numerator $\frac{11}{18} < \frac{5}{8}$ or $\frac{5}{8} > \frac{11}{18}$

$\frac{5}{8} = \frac{45}{72}$ ← Larger Numerator

Example 1 Place the correct symbol, < or >, between the numbers.

$\frac{5}{12}$ $\frac{7}{18}$

Solution $\frac{5}{12} = \frac{15}{36}$ $\frac{7}{18} = \frac{14}{36}$

$\frac{5}{12} > \frac{7}{18}$

Example 2 Place the correct symbol, < or >, between the numbers.

$\frac{9}{14}$ $\frac{13}{21}$

Your solution

Solution on p. A16

Objective B To simplify expressions containing exponents

Repeated multiplication of the same fraction can be written in two ways:

$$\frac{1}{2} \cdot \frac{1}{2} \cdot \frac{1}{2} \cdot \frac{1}{2} \text{ or } \left(\frac{1}{2}\right)^4 \leftarrow \text{exponent}$$

The exponent indicates how many times the fraction occurs as a factor in the multiplication. The expression $\left(\frac{1}{2}\right)^4$ is in exponential notation.

Example 3 Simplify: $\left(\frac{5}{6}\right)^3 \cdot \left(\frac{3}{5}\right)^2$

Solution $\left(\frac{5}{6}\right)^3 \cdot \left(\frac{3}{5}\right)^2 = \left(\frac{5}{6} \cdot \frac{5}{6} \cdot \frac{5}{6}\right) \cdot \left(\frac{3}{5} \cdot \frac{3}{5}\right)$

$= \dfrac{\cancel{5} \cdot \cancel{5} \cdot 5 \cdot \cancel{3} \cdot \cancel{3}}{2 \cdot \cancel{3} \cdot 2 \cdot \cancel{3} \cdot 2 \cdot 3 \cdot \cancel{5} \cdot \cancel{5}} = \dfrac{5}{24}$

Example 4 Simplify: $\left(\frac{7}{11}\right)^2 \cdot \left(\frac{2}{7}\right)$

Your solution

Solution on p. A16

Objective C To use the Order of Operations Agreement to simplify expressions

The Order of Operations Agreement is used for fractions as well as whole numbers.

Step 1 Do all operations inside parentheses.
Step 2 Simplify any number expressions containing exponents.
Step 3 Do multiplications and divisions as they occur from left to right.
Step 4 Do additions and subtractions as they occur from left to right.

Simplify $\frac{14}{15} - \left(\frac{1}{2}\right)^2 \times \left(\frac{2}{3} + \frac{4}{5}\right)$ by using the Order of Operations Agreement.

$\frac{14}{15} - \left(\frac{1}{2}\right)^2 \times \left(\frac{2}{3} + \frac{4}{5}\right)$ 1. Perform operations in parentheses.

$\frac{14}{15} - \left(\frac{1}{2}\right)^2 \times \frac{22}{15}$ 2. Simplify expressions with exponents.

$\frac{14}{15} - \frac{1}{4} \times \frac{22}{15}$ 3. Do multiplications and divisions as they occur from left to right.

$\frac{14}{15} - \frac{11}{30}$ 4. Do additions and subtractions as they occur from left to right.

$\frac{17}{30}$

One or more of the above steps may not be needed to simplify an expression. In that case, proceed to the next step in the Order of Operations Agreement.

Example 5 Simplify: $\left(\frac{3}{4}\right)^2 \div \left(\frac{3}{8} - \frac{1}{12}\right)$

Solution $\left(\frac{3}{4}\right)^2 \div \left(\frac{3}{8} - \frac{1}{12}\right)$

$\left(\frac{3}{4}\right)^2 \div \left(\frac{7}{24}\right)$

$\frac{9}{16} \div \frac{7}{24} = \frac{27}{14} = 1\frac{13}{14}$

Example 6 Simplify:
$\left(\frac{1}{13}\right)^2 \cdot \left(\frac{1}{4} + \frac{1}{6}\right) \div \frac{5}{13}$

Your solution

Solution on p. A16

2.8 EXERCISES

▶ **Objective A**

Place the correct symbol, < or >, between the two numbers.

1. $\dfrac{11}{40} \quad \dfrac{19}{40}$
2. $\dfrac{92}{103} \quad \dfrac{19}{103}$
3. $\dfrac{2}{3} \quad \dfrac{5}{7}$
4. $\dfrac{2}{5} \quad \dfrac{3}{8}$

5. $\dfrac{5}{8} \quad \dfrac{7}{12}$
6. $\dfrac{11}{16} \quad \dfrac{17}{24}$
7. $\dfrac{7}{9} \quad \dfrac{11}{12}$
8. $\dfrac{5}{12} \quad \dfrac{7}{15}$

9. $\dfrac{13}{14} \quad \dfrac{19}{21}$
10. $\dfrac{13}{18} \quad \dfrac{7}{12}$
11. $\dfrac{7}{24} \quad \dfrac{11}{30}$
12. $\dfrac{13}{36} \quad \dfrac{19}{48}$

▶ **Objective B**

Simplify.

13. $\left(\dfrac{3}{8}\right)^2$
14. $\left(\dfrac{5}{12}\right)^2$
15. $\left(\dfrac{2}{9}\right)^3$
16. $\left(\dfrac{5}{7}\right)^3$

17. $\left(\dfrac{1}{2}\right) \cdot \left(\dfrac{2}{3}\right)^2$
18. $\left(\dfrac{2}{3}\right) \cdot \left(\dfrac{1}{2}\right)^4$
19. $\left(\dfrac{1}{3}\right)^2 \cdot \left(\dfrac{3}{5}\right)^3$
20. $\left(\dfrac{2}{5}\right)^2 \cdot \left(\dfrac{5}{8}\right)^3$

21. $\left(\dfrac{2}{5}\right)^3 \cdot \left(\dfrac{5}{7}\right)^2$
22. $\left(\dfrac{5}{9}\right)^3 \cdot \left(\dfrac{18}{25}\right)^2$
23. $\left(\dfrac{1}{3}\right)^4 \cdot \left(\dfrac{9}{11}\right)^2$
24. $\left(\dfrac{8}{15}\right)^3 \cdot \left(\dfrac{5}{8}\right)^4$

25. $\left(\dfrac{1}{2}\right)^6 \cdot \left(\dfrac{32}{35}\right)^2$
26. $\left(\dfrac{2}{3}\right)^4 \cdot \left(\dfrac{81}{100}\right)^2$
27. $\left(\dfrac{1}{6}\right) \cdot \left(\dfrac{6}{7}\right)^2 \cdot \left(\dfrac{2}{3}\right)$

28. $\left(\dfrac{2}{7}\right) \cdot \left(\dfrac{7}{8}\right)^2 \cdot \left(\dfrac{8}{9}\right)$
29. $3 \cdot \left(\dfrac{3}{5}\right)^3 \cdot \left(\dfrac{1}{3}\right)^2$
30. $4 \cdot \left(\dfrac{3}{4}\right)^3 \cdot \left(\dfrac{4}{7}\right)^2$

31. $11 \cdot \left(\dfrac{3}{8}\right)^3 \cdot \left(\dfrac{8}{11}\right)^2$
32. $5 \cdot \left(\dfrac{3}{5}\right)^3 \cdot \left(\dfrac{2}{3}\right)^4$
33. $\left(\dfrac{2}{7}\right)^2 \cdot \left(\dfrac{7}{9}\right)^2 \cdot \left(\dfrac{9}{11}\right)^2$

▶ **Objective C**

Simplify using the Order of Operations Agreement.

34. $\dfrac{1}{2} - \dfrac{1}{3} + \dfrac{2}{3}$

35. $\dfrac{2}{5} + \dfrac{3}{10} - \dfrac{2}{3}$

36. $\dfrac{1}{3} \div \dfrac{1}{2} + \dfrac{3}{4}$

37. $\dfrac{3}{5} \div \dfrac{6}{7} + \dfrac{4}{5}$

38. $\dfrac{4}{5} + \dfrac{3}{7} \cdot \dfrac{14}{15}$

39. $\dfrac{2}{3} + \dfrac{5}{8} \cdot \dfrac{16}{35}$

40. $\left(\dfrac{3}{4}\right)^2 - \dfrac{5}{12}$

41. $\left(\dfrac{3}{5}\right)^3 - \dfrac{3}{25}$

42. $\dfrac{5}{6} \cdot \left(\dfrac{2}{3} - \dfrac{1}{6}\right) + \dfrac{7}{18}$

43. $\dfrac{3}{4} \cdot \left(\dfrac{11}{12} - \dfrac{7}{8}\right) + \dfrac{5}{16}$

44. $\dfrac{7}{12} - \left(\dfrac{2}{3}\right)^2 + \dfrac{5}{8}$

45. $\dfrac{11}{16} - \left(\dfrac{3}{4}\right)^2 + \dfrac{7}{12}$

46. $\dfrac{3}{4} \cdot \left(\dfrac{4}{9}\right)^2 + \dfrac{1}{2}$

47. $\dfrac{9}{10} \cdot \left(\dfrac{2}{3}\right)^3 + \dfrac{2}{3}$

48. $\left(\dfrac{1}{2} + \dfrac{3}{4}\right) \div \dfrac{5}{8}$

49. $\left(\dfrac{2}{3} + \dfrac{5}{6}\right) \div \dfrac{5}{9}$

50. $\dfrac{3}{8} \div \left(\dfrac{5}{12} + \dfrac{3}{8}\right)$

51. $\dfrac{7}{12} \div \left(\dfrac{2}{3} + \dfrac{5}{9}\right)$

52. $\left(\dfrac{3}{8}\right)^2 \div \left(\dfrac{3}{7} + \dfrac{3}{14}\right)$

53. $\left(\dfrac{5}{6}\right)^2 \div \left(\dfrac{5}{12} + \dfrac{2}{3}\right)$

54. $\dfrac{2}{5} \div \dfrac{3}{8} \cdot \dfrac{4}{5}$

55. $\dfrac{7}{8} - \left(\dfrac{1}{6} + \dfrac{5}{12}\right) \div \dfrac{7}{8}$

56. $\left(\dfrac{3}{4}\right)^2 \cdot \left(\dfrac{1}{4} + \dfrac{3}{8}\right) \div \dfrac{5}{12}$

57. $\left(\dfrac{2}{3}\right)^2 + \left(\dfrac{1}{2} - \dfrac{1}{3}\right) \div \dfrac{3}{8}$

58. $\left(1\dfrac{1}{3} - \dfrac{5}{6}\right) + \dfrac{7}{8} \div \left(\dfrac{1}{2}\right)^3$

Calculators and Computers

Fractions The program TO ADD OR SUBTRACT FRACTIONS can be found on the Math ACE disk. This program will allow you to practice adding and subtracting fractions.

The program is designed to provide you with a problem. Then, using a pencil and paper, you are to solve the problem. After you have solved the problem, press the RETURN key; the solution will be displayed. Compare this solution with your solution.

Chapter Summary

Key Words The *least common multiple* (LCM) is the smallest common multiple of two or more numbers.

The *greatest common factor* (GCF) is the largest common factor of two or more numbers.

A *fraction* can represent the number of equal parts of a whole.

A *proper fraction* is a fraction less than 1.

A *mixed number* is a number greater than 1 with a whole number part and a fractional part.

An *improper fraction* is a fraction greater than or equal to 1.

Equal fractions with different denominators are called *equivalent fractions*.

A fraction is in *simplest form* when there are no common factors in the numerator and the denominator.

The *reciprocal* of a fraction is the fraction with the numerator and denominator interchanged.

Inverting is the process of finding the reciprocal of a fraction.

Essential Rules ***Addition of Fractions with Like Denominators***

To add fractions with like denominators, add the numerators and place the sum over the common denominator.

Addition of Fractions with Unlike Denominators

To add fractions with unlike denominators, first rewrite the fractions as equivalent fractions with the same denominator. Then add the numerators and place the sum over the common denominator.

Subtraction of Fractions with Like Denominators

To subtract fractions with like denominators, subtract the numerators and place the difference over the common denominator.

Subtraction of Fractions with Unlike Denominators

To subtract fractions with unlike denominators, rewrite the fractions as equivalent fractions with the same denominator. Then subtract the numerators and place the difference over the common denominator.

Multiplication of Fractions

To multiply two fractions, multiply the numerators and place the product over the product of the denominators.

Division of Fractions

To divide two fractions, multiply by the reciprocal of the divisor.

Chapter Review

SECTION 2.1

1. Find the LCM of 18 and 12.
2. Find the LCM of 18 and 27.
3. Find the GCF of 20 and 48.
4. Find the GCF of 15 and 25.

SECTION 2.2

5. Express the shaded portion of the circles as an improper fraction.

6. Express the shaded portion of the circles as a mixed number.

7. Write $\frac{17}{5}$ as a mixed number.
8. Write $2\frac{5}{7}$ as an improper fraction.

SECTION 2.3

9. Write an equivalent fraction with the given denominator.
$\frac{2}{3} = \frac{}{36}$

10. Write an equivalent fraction with the given denominator.
$\frac{8}{11} = \frac{}{44}$

11. Write $\frac{30}{45}$ in simplest form.
12. Write $\frac{16}{44}$ in simplest form.

SECTION 2.4

13. Add: $\frac{3}{8} + \frac{5}{8} + \frac{1}{8}$
14. Add: $\frac{2}{3} + \frac{5}{6} + \frac{2}{9}$
15. Add: $\frac{3}{8} + 1\frac{2}{3} + 3\frac{5}{6}$
16. Add: $4\frac{4}{9} + 2\frac{1}{6} + 11\frac{17}{27}$

17. During 3 months of the rainy season, $5\frac{7}{8}$, $6\frac{2}{3}$, and $8\frac{3}{4}$ inches of rain fell. Find the total rainfall for the 3 months.

SECTION 2.5

18. Subtract: $\frac{11}{18} - \frac{5}{18}$
19. Subtract: $\frac{17}{24} - \frac{3}{16}$

20. Subtract: $18\frac{1}{6}$
    $- 3\frac{5}{7}$

21. Subtract: 16
    $- 5\frac{7}{8}$

22. A 15-mile race has three checkpoints. The first checkpoint is $4\frac{1}{2}$ miles from the starting point. The second checkpoint is $5\frac{3}{4}$ miles from the first checkpoint. How many miles is the second checkpoint from the finish line?

SECTION 2.6

23. Multiply: $\frac{5}{12} \times \frac{4}{25}$

24. Multiply: $\frac{11}{50} \times \frac{25}{44}$

25. Multiply: $2\frac{1}{3} \times 3\frac{7}{8}$

26. Multiply: $2\frac{1}{4} \times 7\frac{1}{3}$

27. A compact car gets 36 miles on each gallon of gasoline. How many miles can the car travel on $6\frac{3}{4}$ gallons of gasoline?

SECTION 2.7

28. Divide: $\frac{5}{6} \div \frac{5}{12}$

29. Divide: $\frac{15}{28} \div \frac{5}{7}$

30. Divide: $1\frac{1}{3} \div \frac{2}{3}$

31. Divide: $8\frac{2}{3} \div 2\frac{3}{5}$

32. A home building contractor bought $4\frac{2}{3}$ acres for $168,000. What was the cost of each acre?

SECTION 2.8

33. Place the correct symbol, < or >, between the two numbers.
    $\frac{11}{18}$ $\frac{17}{24}$

34. Simplify: $\left(\frac{3}{4}\right)^3 \cdot \frac{20}{27}$

35. Simplify: $\frac{2}{7}\left(\frac{5}{8} - \frac{1}{3}\right) \div \frac{3}{5}$

36. Simplify: $\left(\frac{4}{5} - \frac{2}{3}\right)^2 \div \frac{4}{15}$

Chapter Test

1. Find the LCM of 24 and 40.

2. Find the GCF of 24 and 80.

3. Express the shaded portion of the circles as an improper fraction.

4. Write $\frac{18}{5}$ as a mixed number.

5. Write $9\frac{4}{5}$ as an improper fraction.

6. Write an equivalent fraction with the given denominator.
   $$\frac{5}{8} = \frac{}{72}$$

7. Write $\frac{40}{64}$ in simplest form.

8. Add: $\frac{7}{12} + \frac{11}{12} + \frac{5}{12}$

9. Add: $\frac{5}{6}$

   $\frac{7}{9}$

   $+\frac{1}{15}$

10. Add: $12\frac{5}{12}$

    $+\ 9\frac{17}{20}$

11. The rainfall for a 3-month period was $11\frac{1}{2}$ inches, $7\frac{5}{8}$ inches, and $2\frac{1}{3}$ inches. Find the total rainfall for the 3 months.

12. Subtract: $\frac{17}{24} - \frac{11}{24}$

13. Subtract: $\frac{9}{16}$

    $-\frac{5}{12}$

14. Subtract: $23\frac{1}{8}$

    $-\ 9\frac{9}{44}$

15. An investor bought 100 shares of a utility stock at $24\frac{1}{2}$. The stock gained $5\frac{5}{8}$ during the first month of ownership and lost $2\frac{1}{4}$ during the second month. Find the value of 1 share of the utility stock at the end of the second month.

16. Multiply: $\frac{9}{11} \times \frac{44}{81}$

17. Multiply: $5\frac{2}{3} \times 1\frac{7}{17}$

18. An electrician earns $120 for each day worked. What is the total of the electrician's earnings for working $3\frac{1}{2}$ days?

19. Divide: $\frac{5}{9} \div \frac{7}{18}$

20. Divide: $6\frac{2}{3} \div 3\frac{1}{6}$

21. An investor bought $7\frac{1}{4}$ acres of land for a housing project. One and three-fourths acres were set aside for a park, and the remaining land was developed into $\frac{1}{2}$-acre lots. How many lots were available for sale?

22. Place the correct symbol, < or >, between the two numbers.
$\frac{3}{8}$ $\frac{5}{12}$

23. Simplify: $\left(\frac{2}{3}\right)^4 \cdot \frac{27}{32}$

24. Simplify: $\left(\frac{3}{4}\right)^2 \div \left(\frac{2}{3} + \frac{5}{6}\right) - \frac{1}{12}$

25. Simplify: $\left(\frac{1}{4}\right)^3 \div \left(\frac{1}{8}\right)^2 - \frac{1}{6}$

Cumulative Review

1. Round 290,496 to the nearest thousand.

2. Subtract: 390,047
   $-$ 98,769

3. Multiply: 926
   $\times$ 79

4. Divide: $57\overline{)30{,}792}$

5. Simplify: $4 \cdot (6 - 3) \div 6 - 1$

6. Find the prime factorization of 44.

7. Find the LCM of 30 and 42.

8. Find the GCF of 60 and 80.

9. Write $7\frac{2}{3}$ as an improper fraction.

10. Write $\frac{25}{4}$ as a mixed number.

11. Write an equivalent fraction with the given denominator.
    $\frac{5}{16} = \frac{}{48}$

12. Write $\frac{24}{60}$ in simplest form.

13. Add: $\frac{7}{12} + \frac{9}{16}$

14. Add: $3\frac{7}{8}$
    $7\frac{5}{12}$
    $+2\frac{15}{16}$

15. Subtract: $\frac{11}{12} - \frac{3}{8}$

16. Subtract: $5\frac{1}{6}$
    $-3\frac{7}{18}$

17. Multiply: $\frac{3}{8} \times \frac{14}{15}$

18. Multiply: $3\frac{1}{8} \times 2\frac{2}{5}$

19. Divide: $\frac{7}{16} \div \frac{5}{12}$

20. Divide: $6\frac{1}{8} \div 2\frac{1}{3}$

21. Simplify: $\left(\frac{1}{2}\right)^3 \cdot \frac{8}{9}$

22. Simplify: $\left(\frac{1}{2} + \frac{1}{3}\right) \div \left(\frac{2}{5}\right)^2$

23. A student had $1359 in a checking account. During the week, the student wrote checks of $128, $54, and $315. Find the amount in the checking account at the end of the week.

24. The tickets for a movie were $5 for an adult and $2 for a student. Find the total income from the sale of 87 adult tickets and 135 student tickets.

25. Find the total weight of three packages that weigh $1\frac{1}{2}$ pounds, $7\frac{7}{8}$ pounds, and $2\frac{2}{3}$ pounds.

26. A board $2\frac{5}{8}$ feet long is cut from a board $7\frac{1}{3}$ feet long. What is the length of the remaining piece?

27. A car travels 27 miles on each gallon of gasoline. How many miles can the car travel on $8\frac{1}{3}$ gallons of gasoline?

28. A contractor purchased $10\frac{1}{3}$ acres of land to build a housing development. The contractor donated 2 acres for a park. How many $\frac{1}{3}$-acre parcels can be sold from the remaining land?

3 Decimals

OBJECTIVES

- To write decimals in standard form and in words
- To round a decimal to a given place value
- To add decimals
- To solve application problems
- To subtract decimals
- To solve application problems
- To multiply decimals
- To solve application problems
- To divide decimals
- To solve application problems
- To convert fractions to decimals
- To convert decimals to fractions
- To identify the order relation between two decimals or between a decimal and a fraction

Decimal Fractions

How would you like to add $\frac{37,544}{23,465} + \frac{5184}{3456}$? These two fractions are very cumbersome, and it would take even a mathematician some time to get the answer.

Well, around 1550, help was on the way with the publication of a book called *La Disme* ("The Tenth"), which urged the use of decimal fractions. A decimal fraction is one in which the denominator is 10, 100, 1000, 10,000, and so on.

This book suggested that all whole numbers were "units" and when written would end with the symbol ⓪. For example, the number 294⓪ would be the number two hundred ninety-four. This is very much like the way numbers are currently written (except for the ⓪).

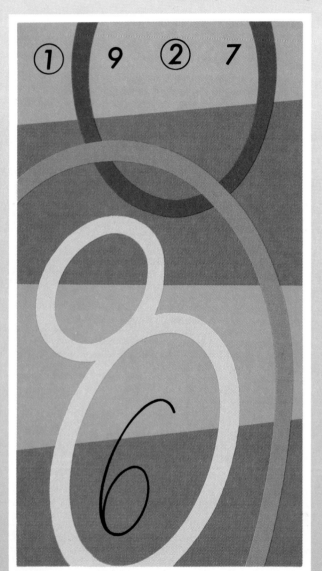

For a fraction between 0 and 1, a unit was broken down into parts called "primes." The fraction three-tenths would be written

$\frac{3}{10} = 3①$ The ① was used to mean the end of the primes, or what are now called tenths.

Each prime was further broken down to "seconds," and each second was broken down to "thirds," and so on. The primes ended with ①, the seconds ended with ②, and the thirds ended with ③.

Examples of these numbers in the old notation and our modern fraction notation are shown below.

$\frac{37}{100} = 3①7②$ $\frac{257}{1000} = 2①5②7③$

After completing this chapter, you might come back to the problem in the first line. Use decimals instead of fractions to find the answer. The answer is 3.1.

SECTION 3.1   Introduction to Decimals

Objective A   To write decimals in standard form and in words

The smallest human bone is found in the middle ear and measures 0.135 inch in length. The number 0.135 is in **decimal notation.**

Note the relationship between fractions and numbers written in decimal notation.

Three-tenths Three-hundredths Three-thousandths

$\frac{3}{10} = 0.\underline{3}$ $\frac{3}{100} = 0.\underline{03}$ $\frac{3}{1000} = 0.\underline{003}$

1 zero 1 decimal place   2 zeros 2 decimal places   3 zeros 3 decimal places

A number written in decimal notation has three parts.

351	.	7089
Whole number part	Decimal point	Decimal part

A number written in decimal notation is often called simply a **decimal.** The position of a digit in a decimal determines the digit's place value.

In the decimal 351.7089, the position of the digit 9 determines that its place value is ten-thousandths.

When writing a decimal in words, write the decimal part as if it were a whole number; then name the place value of the last digit.

0.6481   Six thousand four hundred eighty-one ten-thousandths
549.238   Five hundred forty-nine and two hundred thirty-eight thousandths (The decimal point is read as "and.")

To write a decimal in standard form, zeros may have to be inserted after the decimal point so that the last digit is in the given place-value position.

Five and thirty-eight <u>hundredths</u>

   8 is in the hundredths' place.   5.3<u>8</u>

Nineteen and four <u>thousandths</u>

   4 is in the thousandths' place.   19.00<u>4</u>
   Insert 2 zeros so that 4 is in the
   thousandths place.

Seventy-one <u>ten-thousandths</u>

   1 is in the ten-thousandths'   0.007<u>1</u>
   place. Insert 2 zeros so that 1 is
   in the ten-thousandths place.

Example 1 Write 307.4027 in words.

Solution Three hundred seven and four thousand twenty-seven ten-thousandths

Example 2 Write 209.05838 in words.

Your solution

Example 3 Write six hundred seven and seven hundred eight hundred-thousandths in standard form.

Solution 607.00708

Example 4 Write forty-two thousand and two hundred seven millionths in standard form.

Your solution

Solutions on p. A16

Objective B To round a decimal to a given place value

Rounding decimals is similar to rounding whole numbers except that the digits to the right of the given place value are dropped instead of being replaced by zeros.

If the digit to the right of the given place value is less than 5, that digit and all digits to the right are dropped. If the digit to the right of the given place value is greater than or equal to 5, increase the given place value by 1 and drop all digits to its right.

Round 26.3799 to the nearest hundredth.

⎡—Given place value
26.3799
        ⎣—9 > 5   Increase 7 by one and drop all digits to the right of 7.

26.3799 rounded to the nearest hundredth is 26.38.

Example 5 Round 0.39275 to the nearest ten-thousandth.

Solution
       ⎡—Given place value
0.39275
        ⎣—5 = 5

0.3928

Example 6 Round 4.349254 to the nearest hundredth.

Your solution

Example 7 Round 42.0237412 to the nearest hundred-thousandth.

Solution
           ⎡—Given place value
42.0237412
            ⎣—1 < 5

42.02374

Example 8 Round 3.290532 to the nearest hundred-thousandth.

Your solution

Solutions on p. A16

3.1 EXERCISES

▶ **Objective A**

Write each decimal in words.

1. 0.27
2. 0.92
3. 1.005
4. 3.067

5. 36.4
6. 59.7
7. 0.00035
8. 0.00092

9. 32.107
10. 65.206
11. 12.1294
12. 26.3297

13. 10.007
14. 20.009
15. 52.00095
16. 64.00037

17. 0.0293
18. 0.0717
19. 6.324
20. 8.916

21. 276.3297
22. 418.3115
23. 216.0729
24. 976.0317

25. 4625.0379
26. 2986.0925
27. 1.00001
28. 3.00003

Write each decimal in standard form.

29. Seven hundred sixty-two thousandths
30. Two hundred ninety-five thousandths

31. Sixty-two millionths
32. Forty-one millionths

33. Eight and three hundred four ten-thousandths
34. Four and nine hundred seven ten-thousandths

Write each decimal in standard form.

35. Three hundred four and seven hundredths

36. Eight hundred ninety-six and four hundred seven thousandths

37. Three hundred sixty-two and forty-eight thousandths

38. Seven hundred eighty-four and eighty-four thousandths

39. Three thousand forty-eight and two thousand two ten-thousandths

40. Seven thousand sixty-one and nine thousand one ten-thousandths

41. Twenty-eight and four thousand three hundred seventy-five hundred-thousandths

42. Sixty-seven and eight thousand seven hundred twenty-three hundred-thousandths

43. One thousand three hundred eight and forty-five ten-thousandths

44. Twenty-seven thousand eight and one thousand five hundred-thousandths

▶ Objective B

Round each decimal to the given place value.

45. 7.359   Tenths
46. 6.405   Tenths
47. 23.009   Tenths

48. 89.19204   Tenths
49. 22.68259   Hundredths
50. 16.30963   Hundredths

51. 480.325   Hundredths
52. 670.974   Hundredths
53. 1.03925   Thousandths

54. 7.072854   Thousandths
55. 1946.3745   Thousandths
56. 62.009435   Thousandths

57. 0.029876   Ten-thousandths
58. 0.012346   Ten-thousandths

59. 1.702596   Nearest whole number
60. 2.079239   Hundred-thousandths

61. 0.0102903   Millionths
62. 0.1009754   Millionths

SECTION 3.2 Addition of Decimals

Objective A To add decimals

To add decimals, write the numbers so that the decimal points are on a vertical line. Add as for whole numbers, and write the decimal point in the sum directly below the decimal points in the addends.

Add: 0.237 + 4.9 + 27.32

Tens	Ones	.	Tenths	Hundredths	Thousandths
1	1	.			
	0	.	2	3	7
	4	.	9		
+ 2	7	.	3	2	
3	2	.	4	5	7

Note that by placing the decimal points on a vertical line, digits of the same place value are added.

Example 1 Add:
42.3 + 162.903 + 65.0729

Solution
```
   1 1 1
    42.3
   162.903
 +  65.0729
   270.2759
```

Example 2 Add:
4.62 + 27.9 + 0.62054

Your solution

Example 3 Add: 0.83 + 7.942 + 15

Solution
```
   1 1
    0.83
    7.942
 + 15.
   23.772
```

Example 4 Add: 6.05 + 12 + 0.374

Your solution

Solutions on p. A16

ESTIMATION

Estimating the Sum of Two or More Decimals

Estimate and then use your calculator to find 23.037 + 16.7892.

To estimate the sum of two or more numbers, round each number to the same place value. In this case we will round to the nearest whole number. Then add. The estimated answer is 40.

```
  23.037  ≈    23
+ 16.7892 ≈  + 17
                40
```

Now use your calculator to find the exact result. The exact answer is 39.8262.

`23.037 [+] 16.7892 [=] 39.8262`

Objective B To solve application problems

Example 5

A student bought 10.5, 8.7, and 9.2 gallons of gasoline in 1 month. How much gas did the student buy during the month?

Strategy

To find the total amount of gasoline purchased, add the three amounts (10.5, 8.7, and 9.2)

```
  10.5
   8.7
+  9.2
  28.4
```

The student bought 28.4 gallons of gasoline during the month.

Example 6

A student spent $850.65 for tuition, $282.97 for books, $650 for a room, and $835 for a meal ticket. Find the total cost of the four items.

Your strategy

Your solution

Example 7

A food server earned a salary of $138.50 for working 5 days this week. The food server also received $22.92, $15.80, $19.65, $39.20, and $27.70 in tips during the 5 days. Find the total income for the week.

Strategy

To find the total income, add the tips ($22.92, $15.80, $19.65, $39.20, and $27.70) to the salary ($138.50).

Solution

```
$138.50
  22.92
  15.80
  19.65
  39.20
+ 27.70
$263.77
```

The food server's total income for the week was $263.77.

Example 8

An insurance executive earns a salary of $425 every 4 weeks. During the past 4-week period, the executive received commissions of $485.60, $599.46, $326.75, and $725.42. Find the executive's total income for the past 4-week period.

Your strategy

Your solution

Solutions on p. A16

3.2 EXERCISES

Objective A

Add.

1. 16.008 + 2.0385 + 132.06
2. 17.32 + 1.0579 + 16.5
3. 1.792 + 67 + 27.0526

4. 8.772 + 1.09 + 26.5027
5. 3.02 + 62.7 + 3.924
6. 9.06 + 4.976 + 59.6

7. 82.006 + 9.95 + 0.927
8. 0.826 + 8.76 + 79.005
9. 4.307 + 99.82 + 9.078

10. 2.543 + 7.906 + 2.07 + 1.4092
11. 82.907 + 0.0039 + 234.78308
12. 0.0072 + 96 + 235.09873

13. 0.3
    +0.07

14. 0.29
    +0.4

15. 1.007
    +2.1

16. 7.3
    +9.005

17. 4.9257
    27.05
    + 9.0063

18. 8.72
    99.073
    + 2.9736

19. 62.4
    9.827
    +692.44

20. 8
    89.43
    + 7.0659

Estimate by rounding to the nearest whole number. Then use your calculator to add.

21. 342.42
    89.625
    +176.2

22. 219.9
    0.872
    + 13.42

23. 823.9
    82.65
    + 46.923

24. 678.92
    97.6
    + 5.423

Estimate by rounding to the nearest tenths. Then use your calculator to add.

25. .23
    .39
    +.85

26. .42
    .355
    +.84

27. .8543
    .8772
    +.8591

28. .8972
    .1324
    +.3096

▶ Objective B   **Application Problems**

29. A sales executive received a monthly salary of $655.40 and sales commissions during the month of $1436.78 and $617.32. Find the the sales executive's monthly income.

30. A family receives an electric bill of $79.42, a gas bill of $44.78, and a garbage collection bill of $17.53. Find the total bill for the three services.

31. Find the length of the shaft.

32. Find the length of the shaft.

33. A city's rainfall for the three summer months was 7.33 inches, 12.93 inches, and 10.15 inches. Find the total rainfall for the 3 months.

34. A commuter used 12.4 gallons of gas the first week, 9.8 gallons the second week, 13.5 gallons the third week, and 11.2 gallons the fourth week. Find the total amount of gas used for the 4 weeks.

35. The odometer on a family's car reads 1817.4 miles. The car was driven 8.2 miles on Friday, 79.3 miles on Saturday, and 114.8 miles on Sunday.
    a. How many miles was the car driven during the 3 days?
    b. Find the odometer reading at the end of the 3 days.

36. On a short minivacation, you spent $7.46, $9.93, and $6.54 for gasoline. The motel bills were $32.60 and $45.18.
    a. Find the total amount spent for gasoline.
    b. Find the total amount spent for gasoline and the motel.

37. A food server earns a salary of $178.50 for working a 5-day week. In 1 week, the food server received $27.12, $18.50, $34.10, $22.75, and $29.80 in tips. Find the total income for the week.

38. You have $2142.07 in your checking account. You make deposits of $214.96, $31.42, $1092.43, and $99.27. Find the amount in your checking account after you make the deposits.

39. The average number of points scored per game by each of 10 members of a professional basketball team over an 82-game season is as follows: 31.3, 19.7, 15.4, 11.3, 9.2, 8.7, 6.5, 4.4, 1.2 and 0.3. Find the average number of points the basketball team scored per game.

40. An account executive's commission checks for 6 months are $2542.61, $2509.92, $1977.05, $1656.78, $2946.44, and $2153.88. Find the total commission income for the 6 months.

SECTION 3.3 Subtraction of Decimals

Objective A   To subtract decimals

To subtract decimals, write the numbers so that the decimal points are on a vertical line. Subtract as for whole numbers and write the decimal point in the difference directly below the decimal point in the subtrahend.

Subtract 21.532 − 9.875 and check.

Note that by placing the decimal points on a vertical line, digits of the same place value are subtracted.

$$
\begin{array}{rl}
\text{Check:} & \text{Subtrahend} \quad\quad \overset{1\;1\;\;1\;1}{9.875} \\
& +\,\text{Difference} \quad +11.657 \\
& =\,\text{Minuend} \quad\;\; 21.532
\end{array}
$$

Subtract 4.3 − 1.7942 and check.

If necessary, insert zeros in the minuend before subtracting.

$$
\overset{3\;\;12\;9\;9\;10}{4.3\cancel{0}\cancel{0}\cancel{0}} \quad\text{Check:}\quad \overset{1\;\;\;1\;1\;1}{1.7942}
$$
$$
\begin{array}{r} -1.7942 \\ \hline 2.5058 \end{array} \quad\quad \begin{array}{r} +2.5058 \\ \hline 4.3000 \end{array}
$$

Example 1 Subtract 39.047 − 7.96 and check.

Solution
$$
\overset{8\;\;9\;14}{3\cancel{9}.\cancel{0}\cancel{4}7} \quad\text{Check:}\quad \overset{1\;\;1}{7.96}
$$
$$
\begin{array}{r} -\;\;7.96 \\ \hline 31.087 \end{array} \quad\quad \begin{array}{r} +31.087 \\ \hline 39.047 \end{array}
$$

Example 2 Subtract 72.039 − 8.47 and check.

Your solution

Example 3 Subtract 29 − 9.23 and check.

Solution
$$
\overset{1\;18\;\;9\;10}{2\cancel{9}.\cancel{0}\cancel{0}} \quad\text{Check:}\quad \overset{1\;1\;\;1}{9.23}
$$
$$
\begin{array}{r} -\;\;9.23 \\ \hline 19.77 \end{array} \quad\quad \begin{array}{r} +19.77 \\ \hline 29.00 \end{array}
$$

Example 4 Subtract 35 − 9.67 and check.

Your solution

Example 5 Subtract 1.2 − 0.8235 and check.

Solution
$$
\overset{0\;\;11\;9\;9\;10}{1.2\cancel{0}\cancel{0}\cancel{0}} \quad\text{Check:}\quad \overset{1\;\;\;1\;1\;1}{0.8235}
$$
$$
\begin{array}{r} -0.8235 \\ \hline 0.3765 \end{array} \quad\quad \begin{array}{r} +0.3765 \\ \hline 1.2000 \end{array}
$$

Example 6 Subtract 3.7 − 1.9715 and check.

Your solution

Solutions on p. A17

ESTIMATION
Estimating the Difference Between Two Decimals

Estimate and then use your calculator to find 820.2306 − 475.74815.

To estimate the difference between two numbers, round each number to the same place value. In this case we will round to the nearest tens. Then subtract. The estimated answer is 340.

```
 820.2306  ≈   820
−475.74815 ≈  −480
              ────
               340
```

Now use your calculator to find the exact result. The exact answer is 344.48245.

820.2306 [−] 475.74815 [−] 344.48245

Objective B   To solve application problems

Example 7

You buy a book for $15.87. How much change do you receive from a $20.00 bill?

Strategy

To find the amount of change, subtract the cost of the book ($15.87) from $20.00

Solution
```
 $ 20.00
 − 15.87
 ───────
 $  4.13
```
You receive $4.13 in change.

Example 8

Your breakfast costs $3.85. How much change do you receive from a $5.00 bill?

Your strategy

Your solution

Example 9

You had a balance of $62.41 in your checking account. You then bought a cassette for $8.95, film for $3.17, and a skateboard for $39.77. After paying for these items with a check, how much do you have left in your checking account?

Strategy

To find the new balance:
- Find the total cost of the three items ($8.95 + $3.17 + $39.77).
- Subtract the total cost from the old balance ($62.41).

Solution
```
 $  8.95          $ 62.41
    3.17          − 51.89
 + 39.77          ───────
 ───────          $ 10.52
 $ 51.89  total cost
```
The new balance is $10.52.

Example 10

You had a balance of $2472.69 in your checking account. You then wrote checks for $1025.60, $79.85, and $162.47. Find the new balance in your checking account.

Your strategy

Your solution

Solutions on p. A17

3.3 EXERCISES

Objective A

Subtract and check.

1. 24.037 − 18.41
2. 26.029 − 19.31
3. 123.07 − 9.4273
4. 214 − 7.143

5. 16.5 − 9.7902
6. 13.2 − 8.6205
7. 235.79 − 20.093
8. 463.27 − 40.095

9. 63.005 − 9.1274
10. 23.004 − 7.2175
11. 92 − 19.2909
12. 41.2405 − 25.2709

13. 7.01 − 2.325
14. 8.07 − 5.392
15. 19.0035 − 8.967

16. 0.675
    −0.32

17. 0.924
    −0.91

18. 3
    −1.296

19. 7.507
    −3.419

20. 0.32
    −0.0058

21. 0.78
    −0.0073

22. 3.005
    −1.982

23. 6.007
    −2.734

24. 352.16
    − 90.994

25. 872
    − 80.753

26. 724.32
    − 69

27. 625.46
    − 77.509

28. 362.394
    − 19.4672

29. 421.385
    − 17.5293

30. 19
    −10.372

31. 23.4
    − 0.921

Estimate by rounding to the nearest tens. Then use your calculator to subtract.

32. 620.59
    −132.79

33. 835.07
    −244.82

34. 67.3
    −19.793

35. 84.1
    −48.906

Estimate by rounding to the nearest whole number (nearest ones). Then use your calculator to subtract.

36. 93.079256
   −66.09249

37. 3.7529
   −1.00784

38. 76.53902
   −45.73005

39. 9.07325
   −1.924

▶ Objective B   *Application Problems*

40. A patient has a fever of 102.3° F. Normal temperature is 98.6° F. How many degrees above normal is the patient's temperature?

41. The manager at a fast-food restaurant takes a reading of the cash register tape each hour. At 1:00 P.M. the tape reads $967.54; at 2:00 P.M. the tape reads $1437.15. Find the amount of sales between 1:00 P.M. and 2:00 P.M.

42. A carpenter's plane removes 0.017 inch of wood from a board that is 1.7 inches thick. Find the resulting thickness of the board.

43. A student assistant received a wage increase from $4.75 per hour to $5.24 an hour. Find the per hour raise the assistant received.

44. Find the missing dimension.

45. Find the missing dimension.

46. You earn a salary of $381.24 per week. You have deductions of $30.50 for social security, $18.60 for medical insurance, and $7.18 for union dues.
   a. Find the total amount deducted from your pay.
   b. Find your take-home pay.

47. You had a balance of $1029.74 in your checking account. You then wrote checks for $67.92, $43.10, and $496.34.
   a. Find the total amount of the checks written.
   b. Find the new balance in your checking account.

48. The price of gasoline is $1.22 per gallon after the price rose $.07 one month and $.12 the second month. Find the price of gasoline before the increase in price.

49. Rainfall for the last 3 months of the year was 1.42 inches, 5.39 inches, and 3.55 inches. The normal rainfall for the last 3 months of the year is 11.22 inches. How many inches below normal was the rainfall?

50. A department store had a monthly income of $43,817.62. Salaries amounted to $4214.18, overhead $7124.87, and cost of goods $19,426.44. Find the amount remaining after the three expenses were paid.

51. A telephone company has an income of $4,392,618.58 and pays $692,974.44 in taxes. Find the after-tax income.

SECTION 3.4 Multiplication of Decimals

Objective A To multiply decimals

Decimals are multiplied as if they were whole numbers; then the decimal point is placed in the product. Writing the decimals as fractions shows where to write the decimal point in the product.

$$0.\underline{3} \times 5 = \frac{3}{10} \times \frac{5}{1} = \frac{15}{10} = 1.\underline{5}$$

1 decimal place   1 decimal place

$$0.\underline{3} \times 0.\underline{5} = \frac{3}{10} \times \frac{5}{10} = \frac{15}{100} = 0.\underline{15}$$

1 decimal place   1 decimal place   2 decimal places

$$0.\underline{3} \times 0.\underline{05} = \frac{3}{10} \times \frac{5}{100} = \frac{15}{1000} = 0.\underline{015}$$

1 decimal place   2 decimal places   3 decimal places

To multiply decimals, multiply the numbers as in whole numbers. Write the decimal point in the product so that the number of decimal places in the product is the sum of the decimal places in the factors.

Multiply: 21.4 × 0.36

```
  21.4      1 decimal place
×0.36       2 decimal places
 1284
  642
 7.704      3 decimal places
```

Multiply: 357 × 0.29

```
   357      no decimal places
 ×0.29      2 decimal places
  3213
   714
 103.53     2 decimal places
```

Multiply: 0.037 × 0.08

```
 0.037      3 decimal places
× 0.08      2 decimal places
0.00296     5 decimal places
```

Two zeros must be inserted between the 2 and the decimal point so that there are 5 decimal places in the product.

To multiply a decimal by a power of 10 (10, 100, 1000, . . .), move the decimal point to the right the same number of places as there are zeros in the power of 10.

3.8925 × 1<u>0</u>   = 38.925

   1 zero   1 decimal place

$3.8925 \times 1\underline{00} = 38\overset{\frown}{9.25}$
   2 zeros 2 decimal places

$3.8925 \times 1\underline{000} = 389\overset{\frown}{2.5}$
   3 zeros 3 decimal places

$3.8925 \times 1\underline{0,000} = 38{,}925\overset{\frown}{.}$
   4 zeros 4 decimal places

$3.8925 \times 1\underline{00,000} = 389{,}250\overset{\frown}{.}$ Note that a zero must be inserted
   5 zeros 5 decimal places before the decimal point.

Note that if the power of 10 is written in exponential notation, the exponent indicates how many places to move the decimal point.

$3.8925 \times 10^1 = 38.925$
       1 decimal place

$3.8925 \times 10^2 = 389.25$
       2 decimal places

$3.8925 \times 10^3 = 3892.5$
       3 decimal places

$3.8925 \times 10^4 = 38{,}925.$
       4 decimal places

$3.8925 \times 10^5 = 389{,}250.$
       5 decimal places

Example 1 Multiply: 920×3.7 **Example 2** Multiply: 870×4.6

Solution
```
    920
  × 3.7
  ─────
   644 0
  2760
  ─────
  3404.0
```

                Your solution

Example 3 Multiply: 0.37×0.9 **Example 4** Multiply: 0.28×0.7

Solution
```
  0.37
× 0.9
─────
0.333
```

                Your solution

Solutions on p. A17

Example 5	Multiply: 0.00079 × 0.025	**Example 6**	Multiply: 0.000086 × 0.057
Solution	0.00079 × 0.025 ――――― 395 158 ――――― 0.00001975	**Your solution**	
Example 7	Multiply: 49.093 × 0.079	**Example 8**	Multiply: 79.063 × 0.47
Solution	49.093 × 0.079 ――――― 441837 343651 ――――― 3.878347	**Your solution**	
Example 9	Multiply: 3.69 × 2.07	**Example 10**	Multiply: 4.68 × 6.03
Solution	3.69 ×2.07 ――――― 2583 7380 ――――― 7.6383	**Your solution**	
Example 11	Multiply: 42.07 × 10,000	**Example 12**	Multiply: 6.9 × 1000
Solution	42.07 × 10,000 = 420,700	**Your solution**	
Example 13	Multiply: 3.01×10^3	**Example 14**	Multiply: 4.0273×10^2
Solution	$3.01 \times 10^3 = 3010$	**Your solution**	

Solutions on p. A17

ESTIMATION

Estimating the Product of Two Decimals

Estimate and then use your calculator to find 28.259 × 0.029.

To estimate a product, round each number so that there is one non-zero digit. Then multiply.

$$\begin{array}{r} 28.259 \approx 30 \\ \times\; 0.029 \approx \times 0.03 \\ \hline 0.90 \end{array}$$

The estimated answer is 0.90.

Now use your calculator to find the exact answer.

28.259 ☒ 0.029 ☐ 0.819511

The exact answer is 0.819511

Objective B — To solve application problems

Example 15

It costs $.036 an hour to operate an electric motor. How much does it cost to operate the motor 120 hours?

Strategy

To find the cost of running the motor 120 hours, multiply the hourly cost ($.036) by the number of hours the motor is run (120).

Solution
```
    $.036
  ×  120
    720
    36
  $4.320
```

The cost of running the motor 120 hours is $4.32.

Example 16

The cost of electricity to run a freezer 1 hour is $.035. This month the freezer has run 210 hours. Find the total cost of running the freezer this month.

Your strategy

Your solution

Example 17

A clerk earns a salary of $280 for a 40-hour work week. This week the clerk worked 12 hours of overtime, at a rate of $10.50 for each hour of overtime worked. Find the clerk's total income for the week.

Strategy

To find the clerk's total income for the week:
- Find the overtime pay by multiplying the hourly overtime rate ($10.50) by the number of hours of overtime worked (12).
- Add the overtime pay to the weekly salary ($280).

Solution
```
   $10.50            $280.00
  ×    12          + 126.00
     21 00           $406.00
    105 0
  $126.00 overtime pay
```

The clerk's total income for this week is $406.00.

Example 18

You make a down payment of $175 on a stereo and agree to make payments of $37.18 a month for the next 18 months. Find the total cost of the stereo.

Your strategy

Your solution

Solutions on p. A18

3.4 Exercises

Objective A

Multiply.

1.	0.9 ×0.4	**2.**	0.7 ×0.9	**3.**	0.5 ×0.6	**4.**	0.3 ×0.7	**5.**	0.5 ×0.5
6.	0.7 ×0.7	**7.**	0.9 ×0.5	**8.**	0.2 ×0.6	**9.**	7.7 ×0.9	**10.**	3.4 ×0.4
11.	9.2 ×0.2	**12.**	2.6 ×0.7	**13.**	7.2 ×0.6	**14.**	6.8 ×0.4	**15.**	7.4 ×0.1
16.	3.8 ×0.1	**17.**	7.9 × 5	**18.**	9.3 × 7	**19.**	0.68 × 4	**20.**	0.83 × 9
21.	0.67 × 0.9	**22.**	0.84 × 0.3	**23.**	0.16 × 0.6	**24.**	0.47 × 0.8	**25.**	2.5 ×5.4
26.	3.9 ×1.9	**27.**	8.4 ×9.5	**28.**	7.6 ×5.8	**29.**	0.83 × 5.2	**30.**	0.24 × 2.7
31.	0.46 × 3.9	**32.**	0.78 × 6.8	**33.**	0.2 ×0.3	**34.**	0.3 ×0.3	**35.**	0.24 × 0.3
36.	0.17 × 0.5	**37.**	1.47 × .09	**38.**	6.37 ×0.05	**39.**	8.92 ×0.004	**40.**	6.75 ×0.007

Multiply.

41. 0.49 ×0.16	**42.** 0.38 ×0.21	**43.** 7.6 ×0.01	**44.** 5.1 ×0.01	**45.** 8.62 × 4
46. 5.83 × 7	**47.** 64.5 × 9	**48.** 37.8 × 8	**49.** 2.19 × 9.2	**50.** 1.25 × 5.6
51. 1.85 ×0.023	**52.** 37.8 ×0.052	**53.** 0.478 × 0.37	**54.** 0.526 × 0.22	**55.** 48.3 ×0.0041
56. 67.2 ×0.0086	**57.** 2.437 × 6.1	**58.** 4.237 × 0.54	**59.** 0.413 ×0.0016	**60.** 0.517 ×0.0029
61. 94.73 × 0.57	**62.** 89.23 × 0.62	**63.** 8.005 ×0.067	**64.** 9.032 ×0.019	**65.** 4.29 × 0.1

66. 6.78×0.1 **67.** 5.29×0.4 **68.** 6.78×0.5 **69.** 0.68×0.7

70. 0.56×0.9 **71.** 1.4×0.73 **72.** 6.3×0.37 **73.** 5.2×7.3

74. 7.4×2.9 **75.** 3.8×0.61 **76.** 7.2×0.72 **77.** 0.32×10

78. 6.93×10 **79.** 0.065×100 **80.** 0.039×100 **81.** 6.2856×1000

Multiply.

82. 3.2954×1000 **83.** 3.2×1000 **84.** $0.006 \times 10{,}000$ **85.** $3.57 \times 10{,}000$

86. 8.52×10^1 **87.** 0.63×10^1 **88.** 82.9×10^2

89. 0.039×10^2 **90.** 6.8×10^3 **91.** 4.9×10^4

92. 6.83×10^4 **93.** 0.067×10^2 **94.** 0.052×10^2

Estimate and then use your calculator to multiply.

95. 28.5 × 3.2

96. 86.3 × 4.4

97. 2.38 ×0.44

98. 9.82 ×0.77

99. 0.866 × 4.5

100. 0.239 × 8.2

101. 4.34 ×2.59

102. 6.87 ×9.98

103. 8.434 ×0.044

104. 7.037 ×0.094

105. 28.44 × 1.12

106. 86.57 × 7.33

107. 49.6854 ×39.0672

108. 2.00547 × 9.672

109. 0.00456 ×0.009542

110. 7.00637 × .0128

Objective B Application Problems

111. It costs $.18 per mile to rent a car. Find the cost to rent a car that is driven 114 miles.

112. The cost of operating an electric motor for 1 hour is $.027. How much does it cost to operate the motor 56 hours? (Round to the nearest cent.)

113. Four hundred empty soft drink cans weigh 18.75 pounds. A recycling center pays $.75 per pound for the cans. Find the amount received for the 400 cans (round to the nearest cent).

114. A recycling center pays $.035 per pound for newspapers. Find the amount received from recycling 420 pounds of newspapers.

115. A broker's fee for buying stock is 0.045 times the price of the stock. An investor buys 100 shares of stock at $38.50 per share. Find the broker's fee.

116. A broker's fee for buying stock is 0.052 times the price of the stock. You buy 100 shares of stock at $62.75 per share. Find the broker's fee (round to the nearest cent).

117. You buy a car for $2000 down and payments of $127.50 each month for 36 months.
    a. Find the amount of the payments over the 36 months.
    b. Find the total cost of the car.

118. A nurse earns a salary of $344 for a 40-hour work week. This week the nurse worked 15 hours of overtime at a rate of $12.90 for each hour of overtime worked.
    a. Find the amount of overtime pay.
    b. Find the nurse's total income for the week.

119. A rental car agency charges $12 a day and $.12 per mile for renting a car. You rented a car for 3 days and drove 235 miles. Find the total cost of renting the car.

120. A taxi cost $1.50 and $0.20 for each $\frac{1}{8}$ mile driven. Find the cost of hiring a taxi to get from the airport to the hotel—a distance of 5.5 miles.

121. The escape velocity of a rocket from the earth is 7.4 miles per second. At this speed, how far will the rocket travel in 1 day? (*Hint:* multiply the speed by the number of seconds in 1 day.)

122. The cash price of a house is $104,200. The house can be bought with a 30-year loan with payments of $822.45 a month for 30 years. Find the total amount of the payments.

SECTION 3.5 Division of Decimals

Objective A To divide decimals

To divide decimals, move the decimal point in the divisor to the right to make it a whole number. Move the decimal in the dividend the same number of places to the right. Place the decimal point in the quotient directly over the decimal point in the dividend, and then divide as in whole numbers.

Divide: $3.25 \overline{)15.275}$

$3.25. \overline{)15.27.5}$

Move the decimal point 2 places to the right in the divisor and then in the dividend. Place the decimal point in the quotient.

```
              4.7
    325.) 1527.5
         -1300
           227 5
          -227 5
               0
```

Moving the decimal point the same number of decimal places in the divisor and dividend does not change the value of the quotient because this process is the same as multiplying the numerator and denominator of a fraction by the same number. In the example above,

$$3.25\overline{)15.275} = \frac{15.275}{3.25} = \frac{15.275 \times 100}{3.25 \times 100} = \frac{1527.5}{325} = 325\overline{)1527.5}$$

When dividing decimals, the quotient is usually rounded off to a specified place value, rather than writing the quotient with a remainder.

Divide: $0.3 \overline{)0.56}$
Round to the nearest hundredth.

The division must be carried to the thousandths' place to round the quotient to the nearest hundredth. Therefore, zeros must be inserted in the dividend so that the quotient has a digit in the thousandths' place.

Divide: 57.93 ÷ 3.24 Round to the nearest thousandth.

```
           17.8796  ≈ 17.880
   3.24.) 57.93.0000
         −32 4
          25 53
         −22 68
           2 85 0
          −2 59 2
             25 80
            −22 68
              3 120
             −2 916
                2040
               −1944
```

{ Zeros must be inserted in the dividend so that the quotient has a digit in the ten-thousandths' place.

To divide a decimal by a power of 10 (10, 100, 1000, . . .), move the decimal point to the left the same number of places as there are zeros in the power of 10.

34.65 ÷ 10 = 3.465

    1 zero   1 decimal place

34.65 ÷ 100 = 0.3465

    2 zeros   2 decimal places

34.65 ÷ 1000 = 0.03465

    3 zeros   3 decimal places

Note that a zero must be inserted between the 3 and the decimal point.

34.65 ÷ 10,000 = 0.003465

    4 zeros   4 decimal places

Note that two zeros must be inserted between the 3 and the decimal point.

If the power of 10 is written in exponential notation, the exponent indicates how many places to move the decimal point.

$34.65 \div 10^1 = 3.465$ 1 decimal place

$34.65 \div 10^2 = 0.3465$ 2 decimal places

$34.65 \div 10^3 = 0.03465$ 3 decimal places

$34.65 \div 10^4 = 0.003465$ 4 decimal places

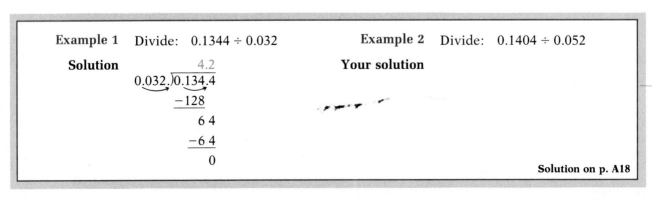

Example 1 Divide: 0.1344 ÷ 0.032

Solution
```
              4.2
   0.032.)0.134.4
          −128
             6 4
            −6 4
               0
```

Example 2 Divide: 0.1404 ÷ 0.052

Your solution

Solution on p. A18

Example 3 Divide: $58.092 \div 82$
Round to the nearest thousandth.

Solution
$$
\begin{array}{r}
0.7084 \approx 0.708 \\
82\overline{)58.0920} \\
-57\ 4 \\
\hline
69 \\
-\ \ 0 \\
\hline
692 \\
-656 \\
\hline
360 \\
-328 \\
\end{array}
$$

Example 4 Divide: $37.042 \div 76$
Round to the nearest thousandth.

Your solution

Example 5 Divide: $420.9 \div 7.06$
Round to the nearest tenth.

Solution
$$
\begin{array}{r}
59.61 \approx 59.6 \\
7.06\overline{)420.90.00} \\
-353\ 0 \\
\hline
67\ 90 \\
-63\ 54 \\
\hline
4\ 36\ 0 \\
-4\ 23\ 6 \\
\hline
12\ 40 \\
-\ \ 7\ 06 \\
\end{array}
$$

Example 6 Divide: $370.2 \div 5.09$
Round to the nearest tenth.

Your solution

Example 7 Divide: $402.75 \div 1000$
Solution $402.75 \div 1000 = 0.40275$

Example 8 Divide: $309.21 \div 10,000$
Your solution

Example 9 Divide: $0.625 \div 10^2$
Solution $0.625 \div 10^2 = 0.00625$

Example 10 Divide: $42.93 \div 10^4$
Your solution

Solutions on p. A18

ESTIMATION

Estimating the Quotient of Two Decimals

Estimate and then use your calculator to find $282.18 \div 0.485$.

To estimate a quotient, round each number so that there is one non-zero digit. Then divide.

$282.18 \div 0.485 \approx$
$300 \div 0.5 = 600$.

The estimated answer is 600.

Now use your calculator to find the exact answer.

$282.18\ \boxed{\div}\ 0.485\ \boxed{=}\ 581.814433$

The exact answer is 581.814433.

Objective B — To solve application problems

Example 11

A retail tire store pays $3.76 excise tax on each tire sold. During 1 month the store paid $300.80 in excise tax. How many tires were sold during the month?

Strategy

To find the number of tires sold, divide the total excise tax paid ($300.80) by the excise tax paid for each tire ($3.76).

Solution

```
              80.
   $3.76.) $300.80.
          −300 8
              00
             − 0
               0
```

The store sold 80 tires during the month.

Example 12

A trucker receives an annual salary of $23,741.40 in 12 equal monthly payments. Find the trucker's monthly income.

Your strategy

Your solution

Example 13

A deluxe microwave oven is on sale for $638.96. You make a down payment of $200 and agree to make 6 equal monthly payments. Find the amount of each monthly payment.

Strategy

To find the amount of each monthly payment:
- Find the balance owed after making the down payment ($638.96 − $200.00).
- Find the monthly payment by dividing the balance by the number of equal monthly payments (6).

Solution

```
   $638.96              $73.16
  −$200.00          6) $438.96
   $438.96 balance    −42
           owed        18
                      −18
                       0 9
                      − 6
                        36
                       −36
                         0
```

The monthly payment is $73.16.

Example 14

A tax of $11.55 is paid on each lamp a store buys. During 1 month the store paid $138.60 in taxes. Each lamp was sold for $95.60. Find the store's total income from the sale of the lamps.

Your strategy

Your solution

Solutions on pp. A18–A19

3.5 EXERCISES

▶ **Objective A**

Divide.

1. $3\overline{)2.46}$
2. $7\overline{)3.71}$
3. $0.8\overline{)3.84}$
4. $0.9\overline{)6.93}$

5. $0.7\overline{)62.3}$
6. $0.4\overline{)52.8}$
7. $0.4\overline{)24}$
8. $0.5\overline{)65}$

9. $7\overline{)0.42}$
10. $3\overline{)0.12}$
11. $0.4\overline{)27.88}$
12. $0.8\overline{)3.312}$

13. $0.7\overline{)59.01}$
14. $0.9\overline{)8.721}$
15. $0.5\overline{)16.15}$
16. $0.8\overline{)77.6}$

17. $0.7\overline{)3.542}$
18. $0.6\overline{)2.436}$
19. $6.3\overline{)8.19}$
20. $3.2\overline{)7.04}$

21. $3.6\overline{)0.396}$
22. $2.7\overline{)0.648}$
23. $6.9\overline{)26.22}$
24. $1.7\overline{)84.66}$

25. $6.1\overline{)64.05}$
26. $3.9\overline{)8.034}$
27. $0.41\overline{)2.501}$

28. $0.59\overline{)1.003}$
29. $0.04\overline{)27.2}$
30. $0.06\overline{)44.4}$

31. $0.08\overline{)23.2}$
32. $0.018\overline{)0.1314}$
33. $0.064\overline{)0.2304}$

Divide. Round to the nearest tenth.

34. 55.62 ÷ 8.8 **35.** 25.43 ÷ 5.4 **36.** 5.427 ÷ 9.5

37. 1.837 ÷ 1.4 **38.** 18.4 ÷ 7.3 **39.** 52.9 ÷ 8.1

40. 0.183 ÷ 0.17 **41.** 0.381 ÷ 0.47 **42.** 6.924 ÷ 0.053

Divide. Round to the nearest hundredth.

43. 4.817 ÷ 16 **44.** 6.467 ÷ 8 **45.** 0.0418 ÷ 0.53

46. 19.08 ÷ 0.45 **47.** 21.792 ÷ 0.96 **48.** 38.665 ÷ 0.95

49. 13.97 ÷ 25.4 **50.** 27.738 ÷ 60.3 **51.** 3.171 ÷ 45.3

Divide. Round to the nearest thousandth.

52. 1.028 ÷ 54 **53.** 6.729 ÷ 27 **54.** 0.0437 ÷ 0.5

55. 75.469 ÷ 77.8 **56.** 34.31 ÷ 95.3 **57.** 0.2695 ÷ 2.67

58. 0.4871 ÷ 4.72 **59.** 0.1142 ÷ 17.2 **60.** 0.2307 ÷ 26.7

Divide. Round to the nearest whole number.

61. 16.5 ÷ 4 **62.** 89.76 ÷ 90 **63.** 1.94 ÷ 0.3

64. 1.0478 ÷ 0.413 **65.** 2.148 ÷ 0.519 **66.** 0.79 ÷ 0.778

67. 3.092 ÷ 0.075 **68.** 392 ÷ 6.9 **69.** 8.729 ÷ 0.075

Divide.

70. 4.07 ÷ 10

71. 0.039 ÷ 10

72. 42.67 ÷ 10

73. 389.7 ÷ 100

74. 1.037 ÷ 100

75. 237.835 ÷ 100

76. 8.295 ÷ 1000

77. 82,547 ÷ 1000

78. 825.37 ÷ 1000

79. $8.35 \div 10^1$

80. $0.32 \div 10^1$

81. $87.65 \div 10^1$

82. $23.627 \div 10^2$

83. $2.954 \div 10^2$

84. $0.0053 \div 10^2$

85. $289.32 \div 10^3$

86. $1.8932 \div 10^3$

87. $0.139 \div 10^3$

Estimate and then use your calculator to divide. Round your calculated answer to the nearest ten-thousandth.

88. 42.42 ÷ 3.8

89. 69.8 ÷ 7.2

90. 389 ÷ 0.44

91. 642 ÷ 0.83

92. 6.394 ÷ 3.5

93. 8.429 ÷ 4.2

94. 1.235 ÷ 0.021

95. 7.456 ÷ 0.072

96. 95.443 ÷ 1.32

97. 423.0925 ÷ 4.0927

98. 1.000523 ÷ 429.07

99. 0.03629 ÷ 0.00054

Objective B Application Problems

100. A professional baseball player got 109 hits out of 423 at bats. Find the player's batting average. (Round to three decimal places.)

101. A high school football player gained 162 yards on 26 carries. Find the numbers of yards gained per carry. (Round to two decimal places.)

102. A park ranger earns $39,440.64 for 12 month's work. How much does the park ranger earn in 1 month?

103. Gasoline tax is $.17 per gallon. Find the number of gallons of gasoline used during a month in which $17.34 was paid in taxes.

104. A family pays $1567.70 a year in car insurance. The insurance is paid in 5 equal payments. Find the amount of each payment.

105. A carpenter is building bookcases that are 3.4 feet long. How many complete shelves can be cut from a 12-foot board?

106. A photography supplies store pays a state tax of $7.26 on each SLR camera it sells for $132.40. This month the total tax paid on SLR cameras was $784.08.
    a. How many cameras were sold during the month?
    b. Find the store's total income from the sale of cameras during the month.

107. You buy a home entertainment center for $1242.58. The down payment is $400, and the balance is to be paid in 18 equal monthly payments.
    a. Find the amount to be paid in monthly payments.
    b. Find the amount of each monthly payment.

108. A tire store pays $353.92 in excise taxes during the month on the sale of 75R-14 steel-belted tires. A tax of $3.16 is paid for each 75R-14 tire sold. Each tire sells for $37.45. Find the tire store's total income from the sale of 75R-14 tires during the month.

109. A tax of $6.92 is paid on each box of electric shavers a store purchases. One month the store paid $256.04 in taxes on electric shavers. Each box of shavers sold for $192.60. Find the store's total income from the sale of the electric shavers.

110. An oil company has 3,541,221,500 shares of stock. The company paid $6,090,990,120 in dividends. Find the dividend for each share of stock. (Round to two decimal places)

111. The total budget for the United States is $900 billion. If each person in the United States were to pay the same amount of taxes, how much would each person pay to raise that amount of money? Assume that there are 270 million people in the United States. (Round to the nearest dollar.)

SECTION 3.6 Comparing and Converting Fractions and Decimals

Objective A To convert fractions to decimals

Every fraction can be written as a decimal. To write a fraction as a decimal, divide the numerator of the fraction by the denominator. The quotient can be rounded to the desired place value.

Convert $\frac{3}{7}$ to a decimal.

$$7\overline{)3.00000} = 0.42857$$

$\frac{3}{7}$ rounded to the nearest hundredth is 0.43.

$\frac{3}{7}$ rounded to the nearest thousandth is 0.429.

$\frac{3}{7}$ rounded to the nearest ten-thousandth is 0.4286

Convert $3\frac{2}{9}$ to a decimal. Round to the nearest thousandth.

$$3\frac{2}{9} = \frac{29}{9} \qquad 9\overline{)29.0000} = 3.2222 \qquad 3\frac{2}{9} \text{ rounded to the nearest thousandth is } 3.222.$$

Example 1 Convert $\frac{3}{8}$ to a decimal. Round to the nearest hundredth.

Solution $8\overline{)3.000} = 0.375 \approx 0.38$

Example 2 Convert $\frac{9}{16}$ to a decimal. Round to the nearest tenth.

Your solution

Example 3 Convert $2\frac{3}{4}$ to a decimal. Round to the nearest tenth.

Solution $2\frac{3}{4} = \frac{11}{4} \qquad 4\overline{)11.00} = 2.75 \approx 2.8$

Example 4 Convert $4\frac{1}{6}$ to a decimal. Round to the nearest hundredth.

Your solution

Solutions on p. A19

Objective B To convert decimals to fractions

To convert a decimal to a fraction, remove the decimal point and place the decimal part over a denominator equal to the place value of the last digit in the decimal.

$0.47 = \dfrac{47}{100}$ (hundredths)

$7.45 = 7\dfrac{45}{100} = 7\dfrac{9}{20}$ (hundredths)

$0.27\overline{5} = \dfrac{275}{1000} = \dfrac{11}{40}$ (thousandths)

$0.16\frac{2}{3} = \dfrac{16\frac{2}{3}}{100} = 16\frac{2}{3} \div 100 = \dfrac{50}{3} \times \dfrac{1}{100} = \dfrac{1}{6}$ (hundredths)

Example 5 Convert 0.82 and 4.75 to fractions.

Solution $0.82 = \frac{82}{100} = \frac{41}{50}$

$4.75 = 4\frac{75}{100} = 4\frac{3}{4}$

Example 6 Convert 0.56 and 5.35 to fractions.

Your solution

Example 7 Convert $0.15\frac{2}{3}$ to a fraction.

Solution $0.15\frac{2}{3} = \frac{15\frac{2}{3}}{100} = 15\frac{2}{3} \div 100$

$= \frac{47}{3} \times \frac{1}{100} = \frac{47}{300}$

Example 8 Convert $0.12\frac{7}{8}$ to a fraction.

Your solution

Solutions on p. A19

Objective C To identify the order relation between two decimals or between a decimal and a fraction

Decimals, like whole numbers and fractions, can be graphed as points on the number line. The number line can be used to show the order of decimals. A decimal that appears to the right of a given number is greater than the given number. A decimal that appears to the left of a given number is less than the given number.

Note that 3, 3.0, and 3.00 represent the same number.

Find the order relation between $\frac{3}{8}$ and 0.38.

$\frac{3}{8} = 0.375$

$0.375 < 0.38$

$\frac{3}{8} < 0.38$

Example 9 Place the correct symbol, < or >, between the numbers.
17.005   17.050

Solution $17.005 < 17.050$

Example 10 Place the correct symbol, < or >, between the numbers.
64.009   64.99

Your solution

Example 11 Place the correct symbol, < or >, between the numbers.
$\frac{5}{16}$ 0.32

Solution $\frac{5}{16} \approx 0.313$

$0.313 < 0.32$

$\frac{5}{16} < 0.32$

Example 12 Place the correct symbol, < or >, between the numbers.
0.63 $\frac{5}{8}$

Your solution

Solutions on p. A19

3.6 EXERCISES

▶ **Objective A**

Convert the fraction to a decimal. Round to the nearest thousandth.

1. $\frac{5}{8}$
2. $\frac{7}{12}$
3. $\frac{2}{3}$
4. $\frac{5}{6}$
5. $\frac{1}{6}$
6. $\frac{7}{8}$

7. $\frac{5}{12}$
8. $\frac{9}{16}$
9. $\frac{7}{4}$
10. $\frac{5}{3}$
11. $1\frac{1}{2}$
12. $2\frac{1}{3}$

13. $\frac{16}{4}$
14. $\frac{36}{9}$
15. $\frac{3}{1000}$
16. $\frac{5}{10}$
17. $7\frac{2}{25}$
18. $16\frac{7}{9}$

19. $37\frac{1}{2}$
20. $87\frac{1}{2}$
21. $\frac{3}{8}$
22. $\frac{11}{16}$
23. $\frac{5}{24}$
24. $\frac{4}{25}$

25. $3\frac{1}{3}$
26. $8\frac{2}{5}$
27. $5\frac{4}{9}$
28. $3\frac{1}{12}$
29. $\frac{5}{16}$
30. $\frac{11}{12}$

31. $2\frac{3}{14}$
32. $1\frac{7}{16}$

▶ **Objective B**

Convert the decimal to a fraction.

33. 0.8
34. 0.4
35. 0.32
36. 0.48
37. 0.125

38. 0.485
39. 1.25
40. 3.75
41. 16.9
42. 17.5

43. 8.4
44. 10.7
45. 8.437
46. 9.279
47. 2.25

Convert the decimal to a fraction.

48. 7.75　　**49.** $0.15\frac{1}{3}$　　**50.** $0.17\frac{2}{3}$　　**51.** $0.87\frac{7}{8}$　　**52.** $0.12\frac{5}{9}$

53. 1.68　　**54.** 7.38　　**55.** 0.045　　**56.** 0.085　　**57.** 16.72

58. 82.32　　**59.** 0.33　　**60.** 0.57　　**61.** $0.33\frac{1}{3}$　　**62.** $0.66\frac{2}{3}$

63. $0.27\frac{3}{7}$　　**64.** $0.29\frac{4}{7}$

▶ Objective C

Place the correct symbol, < or >, between the numbers.

65. 0.15　0.5　　**66.** 0.6　0.45　　**67.** 6.65　6.56　　**68.** 3.89　3.98

69. 2.504　2.054　　**70.** 0.025　0.105　　**71.** $\frac{3}{8}$　0.365　　**72.** $\frac{4}{5}$　0.802

73. $\frac{2}{3}$　0.65　　**74.** 0.85　$\frac{7}{8}$　　**75.** $\frac{5}{9}$　0.55　　**76.** $\frac{7}{12}$　0.58

77. 0.62　$\frac{7}{15}$　　**78.** $\frac{11}{12}$　0.92　　**79.** 0.161　$\frac{1}{7}$　　**80.** 0.623　0.6023

81. 0.86　0.855　　**82.** 0.87　0.087　　**83.** 1.005　0.5　　**84.** 0.033　0.3

85. $\frac{5}{24}$　0.202　　**86.** $\frac{7}{36}$　0.198　　**87.** 0.429　$\frac{4}{9}$　　**88.** 0.835　$\frac{11}{16}$

Calculators and Computers

Using a Calculator to Add or Subtract Fractions

You may recall from your study of this chapter that every fraction can be represented as a decimal. By using a calculator, it is easy to find the decimal equivalent of a fraction. Simply divide the numerator by the denominator.

To find the decimal equivalent of $\frac{7}{30}$, follow these steps:

Enter	Display	Comments
7	7.	
$\div$	7.	
30	30.	
$=$	0.2333333	The answer to $\frac{7}{30}$.

Note: The result displayed in the calculator will only be a decimal *approximation* to many fractions. Using the decimal equivalent of a fraction, it is possible to add or subtract fractions using your calculator. The parentheses keys, $($ and $)$, will be used for this calculation.

To add $\frac{23}{37} + \frac{5}{19}$ using your calculator, follow these steps:

Enter	Display	Comments
(23 $\div$ 37)	0.6216216	The first fraction.
$+$	0.6216216	
(5 $\div$ 19)	0.2631578	The second fraction.
$=$	0.8847795	
$\times$	0.8847795	
(37 $\times$ 19)	703	The product of the two denominators. (Write this number down.)
$=$	621.99991	Round to the nearest whole number (622). The answer is $\frac{622}{703}$.

Chapter Summary

Key Words A number written in *decimal notation* has three parts: a whole number part, a decimal point, and the decimal part.

The position of a digit in a *decimal* determines the digit's *place value*.

Essential Rules **To Write a Decimal in Words**

To write a decimal in words, write the decimal part as if it were a whole number. Then name the place value of the last digit.

To Add Decimals

To add decimals, write the numbers so that the decimal points are on a vertical line. Add as in whole numbers, and place the decimal point in the sum directly below the decimal point in the addends.

To Subtract Decimals

To subtract decimals, place the numbers so that the decimal points are on a vertical line. Subtract as for whole numbers, and write the decimal point in the difference directly below the decimal point in the subtrahend.

To Multiply Decimals

To multiply decimals, multiply the numbers as in whole numbers. Place the decimal point in the product so that the number of decimal places in the product is the sum of the decimal places in the factors.

To Divide Decimals

To divide decimals, move the decimal point in the divisor to make it a whole number. Move the decimal in the dividend the same number of places to the right. Place the decimal point in the quotient directly over the decimal point in the dividend. Then divide as in whole numbers.

To Write a Fraction as a Decimal

To write a fraction as a decimal, divide the numerator of the fraction by the denominator. Round the quotient to the desired number of places.

To Convert a Decimal to a Fraction

To convert a decimal to a fraction, remove the decimal point and place the decimal part over the denominator equal to the place value of the last digit in the decimal.

Chapter Review

SECTION 3.1

1. Write 22.0092 in words.

2. Write 342.37 in words.

3. Write twenty-two and ninety-two ten-thousandths in standard form.

4. Write three and six thousand seven hundred fifty-three hundred-thousandths in standard form.

5. Round 7.93704 to the nearest hundredth.

6. Round 0.05678235 to the nearest hundred-thousandths.

SECTION 3.2

7. Add: 3.42 + 0.794 + 32.5

8. Add: 369.41
         88.3
          9.774
       +366.99

9. You have $237.44 in your checking account. You make deposits of $56.88, $127.40, and $56.30. Find the amount in your checking account after you make the deposits.

SECTION 3.3

10. Subtract: 27.31 − 4.4465

11. Subtract: 7.796
              −2.9175

12. You had a balance of $895.68 in your checking account. You then wrote checks of $145.72 and $88.45. Find the new balance in your checking account.

SECTION 3.4

13. Multiply: 3.08 × 2.9

14. Multiply: 34.79
                   × 0.74

15. The state income tax on the business you own is $560 plus 0.08 times your profit. You made a profit of $63,000 last year. Find the amount of income tax you paid last year.

SECTION 3.5

16. Divide: 3.6515 ÷ 0.067

17. Divide: 0.053)0.349482

18. A car costing $5944.20 is bought with a down payment of $1500 and 36 equal monthly payments. Find the amount of each monthly payment.

SECTION 3.6

19. Convert $\frac{7}{9}$ to a decimal. Round to the nearest thousandths.

20. Convert $2\frac{1}{3}$ to a decimal. Round to the nearest hundredths.

21. Convert 0.375 to a fraction.

22. Convert $0.66\frac{2}{3}$ to a fraction.

23. Place the correct symbol, < or >, between the two numbers.
    0.055   0.1

24. Place the correct symbol, < or >, between the two numbers.
    $\frac{5}{8}$ 0.62

Chapter Test

1. Write 45.0302 in words.

2. Write two hundred nine and seven thousand eighty-six hundred-thousandths in standard form.

3. Round 7.0954625 to the nearest thousandth.

4. Round 0.07395 to the nearest ten-thousandth.

5. Add: 62.3
        4.007
      +189.65

6. Add: 270.93
         97.
          1.976
       + 88.675

7. You received a salary of $363.75, a commission of $954.82, and a bonus of $225. Find your total income.

8. Subtract: 13.027
             − 8.94

9. Subtract: 37.003
             − 9.23674

10. Find the missing dimension.

    4.86 in.
    ?
    6.23 in.

11. Multiply: 1.37
               0.004

12. Multiply: 0.369
    × 6.7

13. A long distance telephone call cost $.85 for the first 3 minutes and $.42 for each additional minute. Find the cost of a 12-minute long distance telephone call.

14. Divide: $0.006\overline{)1.392}$

15. Divide: $0.037\overline{)0.0569}$
    Round to the nearest thousandth.

16. A car is bought for $6392.60, with a down payment of $1250. The balance is paid in 36 monthly payments. Find the amount of each monthly payment.

17. Convert $\frac{9}{13}$ to a decimal. Round to the nearest thousandth.

18. Convert 0.825 to a fraction.

19. Place the correct symbol, < or >, between the two numbers.
    0.66   0.666

20. Place the correct symbol, < or >, between the two numbers.
    $\frac{3}{8}$ 0.35

Cumulative Review

1. Divide: $89\overline{)20{,}932}$

2. Simplify: $2^3 \cdot 4^2$

3. Simplify: $2^2 - (7-3) \div 2 + 1$

4. Find the LCM of 9, 12, and 24.

5. Write $\frac{22}{5}$ as a mixed number.

6. Write $4\frac{5}{8}$ as an improper fraction.

7. Build an equivalent fraction with the given denominator.
$\frac{5}{12} = \frac{}{60}$

8. Add: $\frac{3}{8} + \frac{5}{12} + \frac{9}{16}$

9. Add: $5\frac{7}{12} + 3\frac{7}{18}$

10. Subtract: $9\frac{5}{9} - 3\frac{11}{12}$

11. Multiply: $\frac{9}{16} \times \frac{4}{27}$

12. Multiply: $2\frac{1}{8} \times 4\frac{5}{17}$

13. Divide: $\frac{11}{12} \div \frac{3}{4}$

14. Divide: $2\frac{3}{8} \div 2\frac{1}{2}$

15. Simplify: $\left(\frac{2}{3}\right)^2 \cdot \left(\frac{3}{4}\right)^3$

16. Simplify: $\left(\frac{2}{3}\right)^2 - \left(\frac{2}{3} - \frac{1}{2}\right) + 2$

17. Write 65.0309 in words.

18. Add: 379.006
         27.523
         9.8707
       + 88.2994

19. Subtract: 29.005
                − 7.9286

20. Multiply: 9.074
                × 6.09

21. Divide: $8.09\overline{)17.42963}$ Round to the nearest thousandth.

22. Convert $\frac{11}{15}$ to a decimal. Round to the nearest thousandth.

23. Convert $0.16\frac{2}{3}$ to a fraction.

24. Place the correct symbol, < or >, between the two numbers.
    $\frac{8}{9}$ 0.98

25. An airplane had 204 passengers aboard. During a stop, 97 passengers got off the plane and 127 passengers got on the plane. How many passengers were on the continuing flight?

26. An investor purchased stock at $32\frac{1}{8}$ per share. During the first 2 months of ownership, the stock lost $\$\frac{3}{4}$ and then gained $\$1\frac{1}{2}$. Find the value of each share of stock at the end of the 2 months.

27. You have a checking balance of $814.35. You then write checks for $42.98, $16.43, and $137.56. Find your checking account balance after you write the checks.

28. A machine lathe takes 0.017 inch from a brass bushing that is 1.412 inches thick. Find the resulting thickness of the bushing.

29. The state income tax on your business is $820 plus 0.08 times your profit. You made a profit of $64,860 last year. Find the amount of income tax you paid last year.

30. You bought a camera costing $210.96. The down payment is $20, and the balance is to be paid in 8 equal monthly payments. Find the monthly payment.

4 Ratio and Proportion

OBJECTIVES

- To write the ratio of two quantities in simplest form
- To solve application problems
- To write rates
- To write unit rates
- To solve application problems
- To determine if a proportion is true
- To solve proportions
- To solve application problems

Musical Scales

When a metal wire is stretched tight and then plucked, a sound is heard. Guitars, banjos, and violins are examples of instruments that use this principle to produce music. A piano is another example of this principle, but the sound is produced by the string being struck by a small hammerlike object.

After the string is plucked or struck, the string begins to vibrate. The number of times the string vibrates in 1 second is called the frequency of the vibration, or the pitch. Normally humans can hear vibrations as low as 16 cps (cycles per second) and as high as 20,000 cps. The longer the string, the lower the pitch of the sound; the shorter the string, the higher the pitch of the sound. In fact, a string half as long as another string will vibrate twice as fast.

Vibrating String

Most music that is heard today is based on what is called the chromatic or twelve-tone scale. For this scale, a vibration of 261 cps is called middle C. A string half as long as the string for middle C vibrates twice as fast and produces a musical note one octave higher.

To produce the notes between the two C's, strings are placed that produce the desired pitch. Recall that as the string gets smaller, the pitch increases. A top view of a grand piano illustrates how the strings vary in length.

A well-tempered chromatic scale is one in which the string lengths are chosen so that the ratios of the frequencies of adjacent notes are the same.

$$\frac{C}{C^\#} = \frac{C^\#}{D} = \frac{D}{D^\#} = \frac{D^\#}{E} = \frac{E}{F} = \frac{F}{F^\#} = \frac{F^\#}{G} = \frac{G}{G^\#} = \frac{G^\#}{A} = \frac{A}{A^\#} = \frac{A^\#}{B} = \frac{B}{C}$$

The common ratio for the chromatic scale is approximately $\frac{1}{1.0595}$.

SECTION 4.1 Ratio

Objective A — To write the ratio of two quantities in simplest form

Quantities such as 3 feet, 12 cents, or 9 cars are number quantities written with **units.**

$$\left.\begin{array}{r}3 \text{ feet}\\12 \text{ cents}\\9 \text{ cars}\end{array}\right\}\text{units}$$

These are only some examples of units. Shirts, dollars, trees, miles, and gallons are further examples.

A **ratio** is the comparison of two quantities that have the *same* units. This comparison can be written three different ways:

1. As a fraction
2. As two numbers separated by a colon (:)
3. As two numbers separated by the word TO

The ratio of the lengths of two boards, one 8 feet long and the other 10 feet long, can be written as

1. $\dfrac{8 \text{ feet}}{10 \text{ feet}} = \dfrac{8}{10} = \dfrac{4}{5}$
2. 8 feet : 10 feet = 8 : 10 = 4 : 5
3. 8 feet TO 10 feet = 8 TO 10 = 4 TO 5

A ratio is in **simplest form** when the two numbers do not have a common factor. Notice that the units are not written.

This ratio means that the smaller board is $\frac{4}{5}$ the length of the longer board.

Example 1

Write the comparison $6 to $8 as a ratio in simplest form using a fraction, a colon, and the word TO.

Solution $\dfrac{\$6}{\$8} = \dfrac{6}{8} = \dfrac{3}{4}$

$6 : $8 = 6 : 8 = 3 : 4

$6 TO $8 = 6 TO 8 = 3 TO 4

Example 2

Write the comparison 20 pounds to 24 pounds as a ratio in simplest form using a fraction, a colon, and the word TO.

Your solution

Example 3

Write the comparison 18 quarts to 6 quarts as a ratio in simplest form using a fraction, a colon, and the word TO.

Solution $\dfrac{18 \text{ quarts}}{6 \text{ quarts}} = \dfrac{18}{6} = \dfrac{3}{1}$

18 quarts : 6 quarts
18 : 6 = 3 : 1

18 quarts TO 6 quarts
18 TO 6 = 3 TO 1

Example 4

Write the comparison 64 miles to 8 miles as a ratio in simplest form using a fraction, a colon, and the word TO.

Your solution

Solutions on p. A19

Objective B To solve application problems

Use the table below for Examples 5 and 6.

Board Feet of Wood at a Lumber Store			
Pine	*Ash*	*Oak*	*Cedar*
20,000	18,000	10,000	12,000

Example 5
Find the ratio, as a fraction in simplest form, of the number of board feet of pine to board feet of oak.

Strategy
To find the ratio, write the ratio of board feet of pine (20,000) to board feet of oak (10,000) in simplest form.

Solution
$\dfrac{20{,}000}{10{,}000} = \dfrac{2}{1}$

The ratio is $\dfrac{2}{1}$.

Example 6
Find the ratio, as a fraction in simplest form, of the number of board feet of cedar to board feet of ash.

Your strategy

Your solution

Example 7
The cost of building a patio cover was $250 for labor and $350 for materials. What is the ratio, as a fraction in simplest form, of the cost of materials to the total cost for labor and materials?

Strategy
To find the ratio, write the ratio of the cost of materials ($350) to the total cost ($250 + $350) in simplest form.

Solution
$\dfrac{\$350}{\$250 + \$350} = \dfrac{350}{600} = \dfrac{7}{12}$

The ratio is $\dfrac{7}{12}$.

Example 8
A company spends $20,000 a month for television advertising and $15,000 a month for radio advertising. What is the ratio, as a fraction in simplest form, of the cost of radio advertising to the total cost of radio and television advertising?

Your strategy

Your Solution

Solutions on p. A19

4.1 EXERCISES

▶ **Objective A**

Write the comparison as a ratio in simplest form using a fraction, a colon (:), and the word TO.

1. 3 pints to 15 pints
2. 6 pounds to 8 pounds
3. $40 to $20
4. 10 feet to 2 feet
5. 3 miles to 8 miles
6. 2 hours to 3 hours
7. 37 hours to 24 hours
8. 29 inches to 12 inches
9. 6 minutes to 6 minutes
10. 8 days to 12 days
11. 35 cents to 50 cents
12. 28 inches to 36 inches
13. 30 minutes to 60 minutes
14. 25 cents to 100 cents
15. 32 ounces to 16 ounces
16. 12 quarts to 4 quarts
17. 3 cups to 4 cups
18. 6 years to 7 years
19. $5 to $3
20. 30 yards to 12 yards
21. 12 quarts to 18 quarts
22. $20 to $28
23. 14 days to 7 days
24. 9 feet to 3 feet
25. 12 months to 12 months
26. 75 ounces to 75 ounces
27. 8 tons to 16 tons
28. 8 cents to 40 cents
29. 27 feet to 18 feet
30. 28 days to 20 days

Objective B Application Problems

Solve. Write ratios as fractions in simplest form.

Family Budget						
Housing	Food	Transportation	Taxes	Utilities	Miscellaneous	Total
$800	$400	$300	$350	$150	$400	$2400

31. Use the table to find the ratio of housing cost to total income.

32. Use the table to find the ratio of food to total income.

33. Use the table to find the ratio of utilities cost to food cost.

34. Use the table to find the ratio of transportation cost to housing cost.

35. A school had 400 attend the Friday night performance of a play, and 325 attended the performance on Saturday afternoon. Write the ratio of the number of people attending on Friday to the number of people attending the play Saturday afternoon.

36. One computer prints 80 letters on one line, and a second computer prints 96 letters on one line. Write the ratio of the number of letters per line of the second computer to the number of letters per line of the first computer.

37. A transformer has 40 turns in the primary coil and 480 turns in the secondary coil. State the ratio of the turns in the primary to the number of turns in the secondary coil.

38. A student bought a computer system for $2400; 5 years later the student sold the computer for $900. Find the ratio of the amount the student received for the computer to the cost of the computer.

39. A house had an original value of $90,000 but increased in value to $110,000 in 5 years.
    a. Find the amount of the increase.
    b. What is the ratio of the increase to the original value?

40. A decorator bought one box of ceramic floor tile for $21 and a box of wood tile for $33.
    a. What was the total cost of the box of ceramic tile and the box of wood tile?
    b. What is the ratio of the cost of the wood tile to the total cost?

41. The price of gasoline jumped from $0.98 to $1.26 in 1 year. What is the ratio of the increase in price to the original price?

42. An investor paid $32,000 for an antique car; 5 years later the investor sold the car for $44,000. What is the ratio of the increase in value to the original value the investor paid.

SECTION 4.2   Rates

Objective A   To write rates

A **rate** is a comparison of two quantities that have *different* units. A rate is written as a fraction.

A distance runner ran 26 miles in 4 hours. The distance-to-time rate is written

$$\frac{26 \text{ miles}}{4 \text{ hours}} = \frac{13 \text{ miles}}{2 \text{ hours}}$$

A rate is in **simplest form** when the numbers that form the rate have no common factors. Notice that the units are written as part of the rate.

Example 1   Write "6 roof supports for every 9 feet" as a rate in simplest form.

Solution $\frac{6 \text{ supports}}{9 \text{ feet}} = \frac{2 \text{ supports}}{3 \text{ feet}}$

Example 2   Write "15 pounds of fertilizer for 12 trees" as a rate in simplest form.

Your solution

Solution on p. A20

Objective B   To write unit rates

A **unit rate** is a rate in which the number in the denominator is 1.

$\frac{\$3.25}{1 \text{ pound}}$ or $3.25/pound is read "$3.25 per pound."

To find unit rates, divide the number in the numerator of the rate by the number in the denominator of the rate.

A car traveled 344 miles on 16 gallons of gasoline. To find the miles per gallon (unit rate), divide the numerator of the rate by the denominator of the rate.

$\frac{344 \text{ miles}}{16 \text{ gallons}}$ is the rate.

$16 \overline{)344.0}$ → 21.5   21.5 miles/gallon is the unit rate.

Example 3   Write "300 feet in 8 seconds" as a unit rate.

Solution $\frac{300 \text{ feet}}{8 \text{ seconds}}$ $8\overline{)300.0}$ → 37.5

37.5 feet/second

Example 4   Write "260 miles in 8 hours" as a unit rate.

Your solution

Solution on p. A20

Objective C To solve application problems

Example 5
A grocery store sells 3 grapefruit for 98 cents. What is the cost per grapefruit to the nearest tenth of a cent?

Strategy

To find the cost per grapefruit, divide the cost for 3 grapefruit (98¢) by the number of grapefruit (3).

Solution

$$32.66 \approx 32.7$$
$$3\overline{)98.00}$$

The cost is 32.7¢ per grapefruit.

Example 6
A cyclist rode 47 miles in 3 hours. What is the miles-per-hour rate to the nearest tenth?

Your strategy

Your solution

Example 7
An investor purchased 100 shares of stock for $1500. One year later, the investor sold the 100 shares for $1800. What was the investor's profit per share?

Strategy

To find the investor's profit per share:
- Find the total profit by subtracting the original cost ($1500) from the selling price ($1800).
- Find the profit per share (unit rate) by dividing the total profit by the number of shares of stock (100).

Solution

$1800 $3
−1500   100)$300
$ 300   total profit

The investor's profit per share was $3.

Example 8
A jeweler purchased 5 ounces of gold for $1980. Later, the jeweler sold the 5 ounces for $2075. What was the jeweler's profit per ounce?

Your strategy

Your solution

Solutions on p. A20

4.2 EXERCISES

▶ Objective A

Write as a rate in simplest form.

1. 3 pounds of meat for 4 people

2. 30 ounces in 24 glasses

3. $80 for 12 boards

4. 84 cents for 6 bars of soap

5. 300 miles on 15 gallons

6. 88 feet in 8 seconds

7. 20 children in 8 families

8. 48 leaves on 9 plants

9. 16 gallons in 2 hours

10. 25 ounces in 5 minutes

▶ Objective B

Write as a unit rate.

11. 10 feet in 4 seconds

12. 816 miles in 6 days

13. $1300 earned in 4 weeks

14. $27,000 earned in 12 months

15. 1100 trees planted on 10 acres

16. 3750 words on 15 pages

17. $32.97 earned in 7 hours

18. $315.70 earned in 22 hours

19. 628.8 miles in 12 hours

20. 388.8 miles in 8 hours

21. 344.4 miles on 12.3 gallons of gasoline

22. 409.4 miles on 11.5 gallons of gasoline

23. $349.80 for 212 pounds

24. $11.05 for 3.4 pounds

25. $2231.25 for 3.75 tons

26. $14,259 for 5.25 tons

Objective C Application Problems

27. A student drives 326.6 miles on 11.5 gallons of gas. Find the number of miles driven per gallon of gas.

28. You drive 246.6 miles in 4.5 hours. Find the number of miles driven per hour.

29. The Saturn-5 rocket uses 534,000 gallons of fuel in 2.5 minutes. How much fuel does the rocket use in 1 minute?

30. A ski instructor works 6 months at a ski resort and earns $13,500. What is the ski instructor's wage per month?

31. An investor owning 400 shares of an automobile corporation receives $1160 in dividends. Find the dividend per share.

32. An investor purchased 350 shares of stock for $12,600. What is the cost per share?

33. The total cost of making 5000 compact discs was $26,536.32. Of the discs made, 122 did not meet company standards.
    a. How many compact discs did meet company standards?
    b. What was the cost per disc for those discs that met company standards?

34. A family purchased a 250-pound side of beef for $365.75 and had it packaged. During the packaging, 75 pounds of beef were discarded as waste.
    a. How many pounds of beef were packaged?
    b. What was the cost per pound for the packaged beef?

35. A fruit stand purchased 250 boxes of strawberries for $162.50. All the strawberries were sold for $312.50. What was the profit per box of strawberries?

36. A company's cost to produce 1000 ties was $3250. The company sold the ties for $9490. What was the company's profit per tie?

37. It takes 500 seconds for the light from the sun to reach the earth. The distance from the sun to the earth is 93 million miles. Find the number of miles that light travels per second.

38. A space satellite travels 418,216 miles in 24 hours. How far does the satellite travel per hour? (Round to the nearest tenth.)

SECTION 4.3   Proportions

Objective A   To determine if a proportion is true

A **proportion** is the equality of two ratios or rates.

$$\frac{50 \text{ miles}}{4 \text{ gallons}} = \frac{25 \text{ miles}}{2 \text{ gallons}}$$ Note that the units of the numerators are the same and the units of the denominators are the same.

$$\frac{3}{6} = \frac{1}{2}$$

A proportion is **true** if the fractions are equal when written in lowest terms.

In any true proportion, the "cross products" are equal.

Is $\frac{2}{3} = \frac{8}{12}$ a true proportion?

$$\frac{2}{3} \diagup\!\!\!\diagdown \frac{8}{12} \quad \begin{array}{l} 3 \times 8 = 24 \\ 2 \times 12 = 24 \end{array}$$ Cross products *are* equal.
$\frac{2}{3} = \frac{8}{12}$ is a true proportion.

A proportion is **not true** if the fractions are not equal when reduced to lowest terms.

If the cross products are not equal, then the proportion is not true.

Is $\frac{4}{5} = \frac{8}{9}$ a true proportion?

$$\frac{4}{5} \diagup\!\!\!\diagdown \frac{8}{9} \quad \begin{array}{l} 5 \times 8 = 40 \\ 4 \times 9 = 36 \end{array}$$ Cross products *are not* equal.
$\frac{4}{5} = \frac{8}{9}$ is not a true proportion.

Example 1   Use cross products to determine if $\frac{5}{8} = \frac{10}{16}$ is a true proportion.

Solution $\frac{5}{8} \diagup\!\!\!\diagdown \frac{10}{16} \quad \begin{array}{l} 8 \times 10 = 80 \\ 5 \times 16 = 80 \end{array}$

The proportion is true.

Example 2   Use cross products to determine if $\frac{6}{10} = \frac{9}{15}$ is a true proportion.

Your solution

Example 3   Use cross products to determine if $\frac{62 \text{ miles}}{4 \text{ gallons}} = \frac{33 \text{ miles}}{2 \text{ gallons}}$ is a true proportion.

Solution $\frac{62}{4} \diagup\!\!\!\diagdown \frac{33}{2} \quad \begin{array}{l} 4 \times 33 = 132 \\ 62 \times 2 = 124 \end{array}$

The proportion is not true.

Example 4   Use cross products to determine if $\frac{\$32}{6 \text{ hours}} = \frac{\$90}{8 \text{ hours}}$ is a true proportion.

Your solution

Solutions on p. A20

Objective B To solve proportions

Sometimes one of the numbers in a proportion is unknown. In this case, it is necessary to *solve* the proportion.

To **solve** a proportion, find a number to replace the unknown so that the proportion is true. Do this by following two steps.

Solve $\frac{n}{3} = \frac{6}{9}$ and check.

Find the cross products.
$n \times 9 = 3 \times 6$
$n \times 9 = 18$

To find n, recall the relationship between multiplication and division.

$n \times 9 = 18$ can be written $9\overline{)18}$
Therefore
$n = 18 \div 9 = 2$

Check:

$\frac{2}{3} \times \frac{6}{9}$ → $3 \times 6 = 18$
→ $2 \times 9 = 18$

Example 5

Solve $\frac{n}{12} = \frac{25}{60}$ and check.

Solution

$n \times 60 = 12 \times 25$
$n \times 60 = 300$
$n = 300 \div 60$
$n = 5$
Check:

$\frac{5}{12} \times \frac{25}{60}$ → $12 \times 25 = 300$
→ $5 \times 60 = 300$

Example 6

Solve $\frac{n}{14} = \frac{3}{7}$ and check.

Your solution

Example 7

Solve $\frac{4}{9} = \frac{n}{16}$. Write the answer to the nearest tenth.

Solution

$4 \times 16 = 9 \times n$
$64 = 9 \times n$
$64 \div 9 = n$
$7.1 \approx n$

Note: A rounded answer is an approximation. Therefore the answer to a check will not be exact.

Example 8

Solve $\frac{5}{8} = \frac{n}{20}$. Write the answer to the nearest tenth.

Your solution

Solutions on p. **A20**

Example 9

Solve $\frac{28}{52} = \frac{7}{n}$ and check.

Solution

$28 \times n = 52 \times 7$
$28 \times n = 364$
$n = 364 \div 28$
$n = 13$
Check:

$\frac{28}{52} \bowtie \frac{7}{13} \rightarrow 52 \times 7 = 364$
$\phantom{\frac{28}{52} \bowtie \frac{7}{13}} \rightarrow 28 \times 13 = 364$

Example 10

Solve $\frac{15}{20} = \frac{12}{n}$ and check.

Your solution

Example 11

Solve $\frac{15}{n} = \frac{8}{3}$. Write the answer to the nearest hundredth.

Solution

$15 \times 3 = n \times 8$
$45 = n \times 8$
$45 \div 8 = n$
$5.63 \approx n$

Example 12

Solve $\frac{12}{n} = \frac{7}{4}$. Write the answer to the nearest hundredth.

Your solution

Example 13

Solve $\frac{n}{9} = \frac{3}{1}$ and check.

Solution

$n \times 1 = 9 \times 3$
$n \times 1 = 27$
$n = 27 \div 1$
$n = 27$
Check:

$\frac{27}{9} \bowtie \frac{3}{1} \rightarrow 9 \times 3 = 27$
$\phantom{\frac{27}{9} \bowtie \frac{3}{1}} \rightarrow 27 \times 1 = 27$

Example 14

Solve $\frac{n}{12} = \frac{4}{1}$ and check.

Your solution

Example 15

Solve $\frac{5}{9} = \frac{15}{n}$ and check.

Solution

$5 \times n = 9 \times 15$
$5 \times n = 135$
$n = 135 \div 5$
$n = 27$
Check:

$\frac{5}{9} \bowtie \frac{15}{27} \rightarrow 9 \times 15 = 135$
$\phantom{\frac{5}{9} \bowtie \frac{15}{27}} \rightarrow 5 \times 27 = 135$

Example 16

Solve $\frac{3}{8} = \frac{12}{n}$ and check.

Your solution

Solutions on p. A20

Objective C To solve application problems

Example 17

A mason determines that 9 cement blocks are required for a retaining wall 2 feet long. At this rate, how many cement blocks are required for a retaining wall that is 24 feet long?

Strategy

To find the number of cement blocks for a retaining wall 24 feet long, write and solve a proportion using n to represent the number of blocks required.

Solution

$$\frac{9 \text{ cement blocks}}{2 \text{ feet}} = \frac{n \text{ cement blocks}}{24 \text{ feet}}$$

$9 \times 24 = 2 \times n$
$216 = 2 \times n$
$216 \div 2 = n$
$108 = n$

108 cement blocks are required for a 24-foot retaining wall.

Example 18

Twenty-four jars can be packed in 6 identical boxes. At this rate, how many jars can be packed in 15 boxes?

Your strategy

Your solution

Example 19

The dosage of a certain medication is 2 ounces for every 50 pounds of body weight. How many ounces of this medication are required for a person who weighs 175 pounds?

Strategy

To find the number of ounces of medication for a person weighing 175 pounds, write and solve a proportion using n to represent the number of ounces of medication for a 175-pound person.

Solution $\dfrac{2 \text{ ounces}}{50 \text{ pounds}} = \dfrac{n \text{ ounces}}{175 \text{ pounds}}$

$2 \times 175 = 50 \times n$
$350 = 50 \times n$
$350 \div 50 = n$
$7 = n$

7 ounces of medication are required for a 175-pound person.

Example 20

Three tablespoons of a liquid plant fertilizer are to be added to every 4 gallons of water. How many tablespoons of fertilizer are required for 10 gallons of water?

Your strategy

Your solution

Solutions on pp. A20–A21

4.3 EXERCISES

▶ Objective A

Determine if the proportion is true or not true.

1. $\dfrac{4}{8} = \dfrac{10}{20}$ 2. $\dfrac{39}{48} = \dfrac{13}{16}$ 3. $\dfrac{7}{8} = \dfrac{11}{12}$ 4. $\dfrac{15}{7} = \dfrac{17}{8}$

5. $\dfrac{27}{8} = \dfrac{9}{4}$ 6. $\dfrac{3}{18} = \dfrac{4}{19}$ 7. $\dfrac{45}{135} = \dfrac{3}{9}$ 8. $\dfrac{3}{4} = \dfrac{54}{72}$

9. $\dfrac{16}{3} = \dfrac{48}{9}$ 10. $\dfrac{15}{5} = \dfrac{3}{1}$ 11. $\dfrac{7}{40} = \dfrac{7}{8}$ 12. $\dfrac{9}{7} = \dfrac{6}{5}$

13. $\dfrac{50 \text{ miles}}{2 \text{ gallons}} = \dfrac{25 \text{ miles}}{1 \text{ gallon}}$ 14. $\dfrac{16 \text{ feet}}{10 \text{ seconds}} = \dfrac{24 \text{ feet}}{15 \text{ seconds}}$

15. $\dfrac{6 \text{ minutes}}{5 \text{ cents}} = \dfrac{30 \text{ minutes}}{25 \text{ cents}}$ 16. $\dfrac{16 \text{ pounds}}{12 \text{ days}} = \dfrac{20 \text{ pounds}}{14 \text{ days}}$

17. $\dfrac{\$15}{4 \text{ pounds}} = \dfrac{\$45}{12 \text{ pounds}}$ 18. $\dfrac{270 \text{ trees}}{6 \text{ acres}} = \dfrac{90 \text{ trees}}{2 \text{ acres}}$

19. $\dfrac{300 \text{ feet}}{4 \text{ rolls}} = \dfrac{450 \text{ feet}}{7 \text{ rolls}}$ 20. $\dfrac{1 \text{ gallon}}{4 \text{ quarts}} = \dfrac{7 \text{ gallons}}{28 \text{ quarts}}$

21. $\dfrac{\$65}{5 \text{ days}} = \dfrac{\$26}{2 \text{ days}}$ 22. $\dfrac{80 \text{ miles}}{2 \text{ hours}} = \dfrac{110 \text{ miles}}{3 \text{ hours}}$

23. $\dfrac{7 \text{ tiles}}{4 \text{ feet}} = \dfrac{42 \text{ tiles}}{20 \text{ feet}}$ 24. $\dfrac{15 \text{ feet}}{3 \text{ yards}} = \dfrac{90 \text{ feet}}{18 \text{ yards}}$

Determine if the proportion is true or not true.

25. $\dfrac{60 \text{ gallons}}{4 \text{ hours}} = \dfrac{160 \text{ gallons}}{12 \text{ hours}}$

26. $\dfrac{2500 \text{ words}}{15 \text{ pages}} = \dfrac{5800 \text{ words}}{35 \text{ pages}}$

27. $\dfrac{\$24{,}000}{12 \text{ months}} = \dfrac{\$36{,}000}{18 \text{ months}}$

28. $\dfrac{250 \text{ trees}}{2 \text{ acres}} = \dfrac{1500 \text{ trees}}{12 \text{ acres}}$

▶ **Objective B**

Solve. Round to the nearest hundredth.

29. $\dfrac{n}{4} = \dfrac{6}{8}$

30. $\dfrac{n}{7} = \dfrac{9}{21}$

31. $\dfrac{12}{18} = \dfrac{n}{9}$

32. $\dfrac{7}{21} = \dfrac{35}{n}$

33. $\dfrac{10}{12} = \dfrac{n}{24}$

34. $\dfrac{6}{n} = \dfrac{24}{36}$

35. $\dfrac{3}{n} = \dfrac{15}{10}$

36. $\dfrac{n}{45} = \dfrac{17}{135}$

37. $\dfrac{9}{4} = \dfrac{18}{n}$

38. $\dfrac{7}{15} = \dfrac{21}{n}$

39. $\dfrac{n}{6} = \dfrac{2}{3}$

40. $\dfrac{5}{12} = \dfrac{n}{144}$

41. $\dfrac{n}{5} = \dfrac{7}{8}$

42. $\dfrac{4}{n} = \dfrac{9}{5}$

43. $\dfrac{8}{5} = \dfrac{n}{6}$

44. $\dfrac{n}{11} = \dfrac{32}{4}$

45. $\dfrac{3}{4} = \dfrac{8}{n}$

46. $\dfrac{5}{12} = \dfrac{n}{8}$

47. $\dfrac{36}{20} = \dfrac{12}{n}$

48. $\dfrac{15}{n} = \dfrac{65}{100}$

49. $\dfrac{n}{15} = \dfrac{21}{12}$

50. $\dfrac{40}{n} = \dfrac{15}{8}$

51. $\dfrac{32}{n} = \dfrac{1}{3}$

52. $\dfrac{5}{8} = \dfrac{42}{n}$

53. $\dfrac{18}{11} = \dfrac{16}{n}$

54. $\dfrac{25}{4} = \dfrac{n}{12}$

55. $\dfrac{28}{8} = \dfrac{12}{n}$

56. $\dfrac{n}{30} = \dfrac{65}{120}$

57. $\dfrac{0.3}{5.6} = \dfrac{n}{25}$

58. $\dfrac{1.3}{16} = \dfrac{n}{30}$

59. $\dfrac{0.7}{9.8} = \dfrac{3.6}{n}$

60. $\dfrac{1.9}{7} = \dfrac{13}{n}$

61. $\dfrac{18.6}{23.59} = \dfrac{88.5}{n}$

62. $\dfrac{4065}{699} = \dfrac{n}{3260}$

63. $\dfrac{8.23}{n} = \dfrac{1.025}{7.093}$

64. $\dfrac{n}{4.72} = \dfrac{3827}{69.4}$

Objective C   Application Problems

Solve. Round to the nearest hundredth.

65. A 20-ounce box of raisin bran contains 2400 calories. How many calories are in a 1.4-ounce serving of the cereal?

66. A 10-ounce container of egg noodles contains 700 milligrams of cholesterol. How many milligrams of cholesterol are in a 2-ounce serving of egg noodles?

67. A homeowner uses 2 pounds of fertilizer for every 100 square feet of lawn. At this rate, how many pounds of fertilizer are used on a lawn that measures 2500 square feet?

68. A nursery provides a liquid plant food by adding 1 gallon of water for each 2 ounces of plant food. At this rate, how many ounces of plant food are required for 25 gallons of water?

69. A savings and loan company receives $10.84 each month for every $1000 it loans. At this rate, what is the monthly payment for a $15,000 loan?

70. A life insurance policy cost $6.87 for every $1000 of insurance. At this rate, what is the cost of $25,000 of insurance?

71. The property tax on a $90,000 home is $1800. At this rate, what is the property tax on a home worth $125,000?

72. The sales tax on a $90 purchase is $4.50. At this rate, what is the sales tax on a $300 purchase?

73. A brick wall 20 feet in length contains 1040 bricks. At the same rate, how many bricks would it take to build a wall 48 feet in length?

74. A farmer has 50 acres of wheat that yield 2900 bushels of wheat. At the same rate, find the number of bushels of wheat the farmer can expect from another 240 acres of wheat.

75. The scale on a map is 1 inch equals 12 miles. Find the distance between two cities that are 9.5 inches apart on the map.

76. The scale on the plans for a new house is 1 inch equals 3 feet. Find the length and width of a room that measures 5 inches by 8 inches on the drawing.

77. The dosage for a medication is $\frac{1}{3}$ ounce for every 40 pounds of body weight. At this rate, how many ounces of medication are required for a person who weighs 180 pounds?

Solve. Round to the nearest hundredth.

78. A lawyer on a reducing program has lost 3 pounds in 5 weeks. At the same rate, how long will it take the lawyer to lose 27 pounds?

79. A pre-election survey showed that 2 out of every 3 eligible voters would cast ballots in the county election. At this rate, how many people in a county of 240,000 eligible voters would vote in the election?

80. A paint manufacturer suggests using 1 gallon of paint for every 400 square feet of a wall. At this rate, how many gallons of paint would be required for a room that has 1400 square feet of wall?

81. An automobile recall was based on tests that showed 24 braking defects in 1000 cars. At this rate, how many defects would be found in 40,000 cars?

82. From previous experience, a manufacturer knows that in an average production run of 2000 circuit boards, 60 will be defective. What number of defective circuit boards can be expected from a run of 25,000 circuit boards?

83. You own 240 shares of a computer stock. The company declares a stock split of 5 shares for every 3 owned. How many shares of stock do you own after the stock split?

84. An investor own 50 shares of a utility stock that pays dividends of $125. At this rate, what dividend would the investor receive after buying 300 additional shares of stock in the utility company?

85. A community college estimates that the ratio of student time to administrative time on a computer is 3:2. During the month in which the computer was used 200 hours for administration, how many hours were used by students?

86. A management consulting firm recommends that the ratio of midmanagement salaries to junior management salaries be 5:4. Using this recommendation, what is the yearly midmanagement salary when the junior management yearly salary is $32,000?

87. Seawater contains 3.5 parts of salt for every 100 parts of seawater. At this rate, what is the amount of salt that can be recovered from 50,000 pounds of seawater?

88. A survey indicates that 19 out of 30 eligible voters will vote in the next election. How many votes will be cast out of 528,000 eligible voters?

Calculators and Computers

The Memory Key The memory key on a calculator is useful when calculating with unit rates. The unit rate is stored in the memory of the calculator and then recalled when it is needed for a computation.

The cost of a car loan is an example of a unit rate. A bank charges a certain dollar amount each month for each dollar that is borrowed. For example, to repay a 5-year loan at an annual interest rate of 9.5%, the borrower must pay $0.0210019 each month for each $1 borrowed.

1. Find the monthly payment for an $8500 loan.
2. Find the monthly payment for a $6200 loan.
3. Find the monthly payment for a $7800 loan.

To answer question 1, enter the following on your calculator:

$$0.0210019 \; \boxed{M+} \; 8500 \; \boxed{\times} \; \boxed{MR} \; \boxed{=}$$

The numbers 178.5161 will be displayed on your calculator. The monthly payment is $178.52 when rounded to the nearest cent.

The $\boxed{M+}$ key means "Add to Memory" and the $\boxed{MR}$ key means "Recall from Memory". Another key on your calculator is the $\boxed{MC}$ key, which means "Clear the Memory".

To answer the second question, it is not necessary to enter the payment rate because it is already stored in the calculator's memory. Just enter the following:

$$6200 \; \boxed{\times} \; \boxed{MR} \; \boxed{=}$$

The number 130.2117 will be displayed on your calculator. The monthly payment is $130.21 rounded to the nearest cent.

Enter the following to answer the third question:

$$7800 \; \boxed{\times} \; \boxed{MR} \; \boxed{=}$$

The number 163.8148 will be displayed on your calculator. The monthly payment is $163.81 rounded to the nearest cent.

The advantage of using the memory key is that various monthly payments can be calculated for different loan amounts without the payment rate having to be re-entered each time.

Chapter Summary

Key Words   Quantities such as 8 feet and 60 miles are number quantities written with *units*.

A *ratio* is the comparison of two quantities that have the same units.

A ratio is in *simplest form* when the two numbers that form the ratio have no common factors.

A *rate* is the comparison of two quantities that have different units.

A rate is in *simplest form* when the numbers that form the rate have no common factors.

A *unit rate* is a rate in which the number in the denominator is 1.

A *proportion* is the equality of two ratios or rates.

Essential Rules   **To Find Unit Rates**

To find unit rates, divide the number in the numerator of the rate by the number in the denominator of the rate.

To Solve a Proportion

One of the numbers in a proportion may be unknown. To solve a proportion, find a number to replace the unknown so that the proportion is true.

Ways to Express a Ratio

A ratio can be written three different ways:

1. As a fraction
2. As two numbers separated by a colon (:)
3. As two numbers separated by the word TO

Chapter Review

SECTION 4.1

1. Write the comparison 8 feet to 28 feet as a ratio in simplest form using a fraction, a colon (:), and the word TO.

2. Write the comparison 6 inches to 15 inches as a ratio in simplest form using a fraction, a colon (:), and the word TO.

3. Write the comparison 12 days to 12 days as a ratio in simplest form using a fraction, a colon (:), and the word TO.

4. Write the comparison 32 dollars to 80 dollars as a ratio in simplest form using a fraction, a colon (:), and the word TO.

5. The high temperature during a 24-hour period was 84 degrees and the low temperature was 42 degrees. Write the ratio of the high temperature to the low temperature for the 24-hour period.

6. A house had an original value of $80,000 but increased in value to $120,000 in 2 years. Find the ratio of the increase to the original value.

7. In 5 years the price of a calculator went from $40 to $24. What is the ratio of the decrease in price to the original price?

8. A retail computer store spends $30,000 a year on TV advertising and $12,000 on newspaper advertising. Find the ratio of TV advertising to newspaper advertising.

SECTION 4.2

9. Write "100 miles in 3 hours" as a rate in simplest form.

10. Write "$15 in 4 hours" as a rate in simplest form.

11. Write "250 miles in 4 hours" as a unit rate.

12. Write "$300 earned in 40 hours" as a unit rate.

13. Write $326.4 miles on 12 gallons" as a unit rate.

14. Write "$8.75 for 5 pounds" as a unit rate.

15. An investor purchased 80 shares of stock for $3580. What is the cost per share?

16. You drove 198.8 miles in 3.5 hours. Find the average number of miles driven per hour.

17. A 15-pound turkey costs $10.20. What is the cost per pound?

18. The total cost of manufacturing 1000 radios was $36,600. Of the radios made, 24 did not pass inspection. Find the cost per radio of the radios that did pass inspection.

SECTION 4.3

19. Determine if the proportion is true or not true.
$$\frac{5}{7} = \frac{25}{35}$$

20. Determine if the proportion is true or not true.
$$\frac{2}{9} = \frac{10}{45}$$

21. Determine if the proportion is true or not true.
$$\frac{8}{15} = \frac{32}{60}$$

22. Determine if the proportion is true or not true.
$$\frac{3}{8} = \frac{10}{24}$$

23. Solve the proportion.
$$\frac{n}{8} = \frac{9}{2}$$

24. Solve the proportion.
$$\frac{16}{n} = \frac{4}{17}$$

25. Solve the proportion. Round to hundredths.
$$\frac{24}{11} = \frac{n}{30}$$

26. Solve the proportion. Round to hundredths.
$$\frac{18}{35} = \frac{10}{n}$$

27. A homeowner used 1.5 pounds of fertilizer for every 200 square feet of lawn. How many pounds of fertilizer are used on a lawn that measures 3000 square feet?

28. An insurance policy costs $3.87 for every $1000 of insurance. At this rate, what is the cost of $50,000 of insurance?

29. The property tax on a $45,000 home is $900. At the same rate, what is the property tax on a home valued at $120,000?

30. A brick wall 40 feet in length contains 448 concrete blocks. At the same rate, how many blocks would it take to build a wall that is 120 feet in length?

Chapter Test

1. Write the comparison 12 days to 8 days as a ratio in simplest form using a fraction, a colon (:), and the word TO.

2. Write the comparison 18 feet to 30 feet as a ratio in simplest form using a fraction, a colon (:), and the word TO.

3. Write the comparison 27 dollars to 81 dollars as a ratio in simplest form using a fraction, a colon (:), and the word TO.

4. Write the comparison 40 miles to 240 miles as a ratio in simplest form using a fraction, a colon (:), and the word TO.

5. The average summer temperature in a California desert is 112 degrees. In a city 100 miles away, the average summer temperature is 86 degrees. Write the ratio of the city temperature to the desert temperature.

6. An automobile sales company spends $25,000 each month for television advertising and $40,000 each month for radio advertising. Find the ratio, as a fraction in simplest form, of the cost of radio advertising to the total cost of advertising.

7. Write "18 supports for every 8 feet" as a rate in simplest form.

8. Write "$81 for 12 boards" as a rate in simplest form.

9. Write "$22,036.80 earned in 12 months" as a unit rate.

10. Write "256.2 miles on 8.4 gallons of gas" as a unit rate.

11. 40 feet of lumber costs $69.20. What is the per foot cost of the lumber?

12. A plane travels 2421 miles in 4.5 hours. Find the plane's speed in miles per hour.

13. Determine if the proportion is true or not true.
$$\frac{5}{14} = \frac{25}{70}$$

14. Determine if the proportion is true or not true.
$$\frac{40}{125} = \frac{5}{25}$$

15. Solve the proportion.
$$\frac{n}{18} = \frac{9}{4}$$

16. Solve the proportion.
$$\frac{5}{12} = \frac{60}{n}$$

17. A research scientist estimates that the human body contains 88 pounds of water for every 100 pounds of body weight. At this rate, estimate the number of pounds of water in a college student weighing 150 pounds.

18. The dosage of a medicine is $\frac{1}{4}$ ounce for every 50 pounds of body weight. How many ounces of this medication are required for a person who weighs 175 pounds?

19. The property tax on a house valued at $75,000 is $1500. At the same rate, find the property tax on a house valued at $140,000.

20. Fifty shares of a utility stock has a dividend of $62.50. At the same rate, find the dividend of 500 shares of the utility stock.

Cumulative Review

1. Subtract: 20,095
   −10,937

2. Write $2 \cdot 2 \cdot 2 \cdot 2 \cdot 3 \cdot 3 \cdot 3$ in exponential notation.

3. Simplify $4 - (5 - 2)^2 \div 3 + 2$.

4. Find the prime factorization of 160.

5. Find the LCM of 9, 12, and 18.

6. Find the GCF of 28 and 42.

7. Reduce $\frac{40}{64}$ to simplest form.

8. Add: $3\frac{5}{6} + 4\frac{7}{15}$

9. Subtract: $10\frac{1}{6} - 4\frac{5}{9}$

10. Multiply: $\frac{11}{12} \times 3\frac{1}{11}$

11. Divide: $3\frac{1}{3} \div \frac{5}{7}$

12. Simplify: $\left(\frac{2}{5} + \frac{3}{4}\right) \div \frac{3}{2}$

13. Write 4.0709 in words.

14. Round 2.09762 to the nearest hundredth.

15. Divide: $8.09 \overline{)16.0976}$
    Round to the nearest thousandth.

16. Convert $0.06\frac{2}{3}$ to a fraction.

Cumulative Review

17. Write the comparison 25 miles to 200 miles as a ratio in simplest form.

18. Write "87 cents for 6 bars of soap" as a rate in simplest form.

19. Write "250.5 miles on 7.5 gallons of gas" as a unit rate.

20. Solve $\frac{40}{n} = \frac{160}{17}$. Round to the nearest hundredth.

21. A car traveled 457.6 miles in 8 hours. Find the car's speed in miles per hour.

22. Solve the proportion.
    $\frac{12}{5} = \frac{n}{15}$

23. You had $1024 in your checking account. You then wrote checks for $192 and $88. What is your new checking account balance?

24. A farmer buys a tractor for $22,760. A down payment of $5000 is required. The balance remaining is paid in 48 equal monthly installments. What is the monthly payment?

25. An assignment for a liberal arts student is to read a book containing 175 pages. The student reads $\frac{2}{5}$ of the book during Thanksgiving vacation. How many pages of the assignment remain to be read?

26. A building contractor bought $2\frac{1}{3}$ acres of land for $84,000. What was the cost of each acre?

27. A shopper bought a shirt for $21.79 and a tie for $8.59. A $50 bill was used to pay for the purchases. Find the amount of change.

28. A college baseball player had 42 hits in 155-at-bats. Find the baseball player's batting average. (Round to three decimal places.)

29. A soil conservationist estimates that a river bank is eroding at the rate of 3 inches every 6 months. At this rate, how many inches will be eroded in 50 months?

30. The dosage of a medicine is $\frac{1}{2}$ ounce for every 50 pounds of body weight. How many ounces of this medication are required for a person who weighs 160 pounds?

5 Percents

OBJECTIVES

- To write a percent as a fraction or decimal
- To write a fraction or a decimal as a percent
- To find the amount when the percent and base are given
- To solve application problems
- To find the percent when the base and amount are given
- To solve application problems
- To find the base when the percent and amount are given
- To solve application problems
- To solve percent problems using proportions
- To solve application problems

The Percent Symbol

The idea of using percent dates back many hundreds of years. Percents are used in business for all types of consumer and business loans; in chemistry to measure the percent concentration of an acid; in economics to measure the increases or decreases of the consumer price index (CPI); and in many other areas that affect our daily lives.

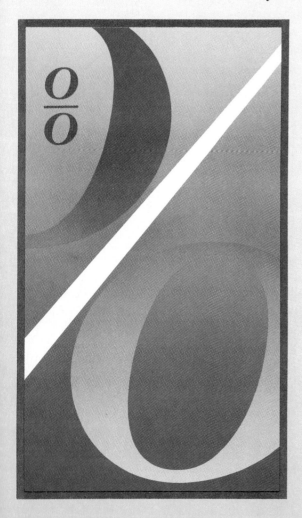

The word percent comes from the Latin phrase *per centum,* which means "by the hundred." The symbol that is used today for percent is %, but this was not always the symbol.

The present symbol apparently is a result of abbreviations for the word "percent."

One abbreviation was p. cent; later, p. 100 and p. $\overset{o}{c}$ were used.

From p. $\overset{o}{c}$, the abbreviation changed to p $\frac{o}{o}$ around the 17th century. This probably was a result of continual writing of p. $\overset{o}{c}$ and the eventual closing of the "c" to make an "o." By the 19th century, the "p" in front of the symbol p$\frac{o}{o}$ was no longer written. The bar that separated the o's became a slash, and the modern symbol % became widely used.

SECTION 5.1 Introduction to Percents

Objective A To write a percent as a fraction or decimal

Percent means "parts of 100." In the figure at the right, there are 100 parts. Because 13 of the 100 parts are shaded, 13% of the figure is shaded.

In most applied problems involving percents, it is necessary to rewrite either a percent as a fraction or a decimal or a fraction or a decimal as a percent.

To write a percent as a fraction, remove the percent sign and multiply by $\frac{1}{100}$.

$$13\% = 13 \times \frac{1}{100} = \frac{13}{100}$$

To write a percent as a decimal, remove the percent sign and multiply by 0.01.

$$13\% = 13 \times 0.01 = 0.13$$

Move the decimal point two places to the left. Then remove the percent sign.

Example 1 Write 120% as a fraction and as a decimal.

Solution
$$120\% = 120 \times \frac{1}{100} = \frac{120}{100}$$
$$= 1\frac{1}{5}$$

$$120\% = 120 \times 0.01 = 1.2$$

Note that percents larger than 100 are greater than 1.

Example 2 Write 125% as a fraction and as a decimal.

Your solution

Example 3 Write $16\frac{2}{3}\%$ as a fraction.

Solution
$$16\frac{2}{3}\% = 16\frac{2}{3} \times \frac{1}{100}$$
$$= \frac{50}{3} \times \frac{1}{100} = \frac{50}{300} = \frac{1}{6}$$

Example 4 Write $33\frac{1}{3}\%$ as a fraction.

Your solution

Example 5 Write 0.5% as a decimal.

Solution $0.5\% = 0.5 \times 0.01 = 0.005$

Example 6 Write 0.25% as a decimal.

Your solution

Solutions on p. A21

Objective B To write a fraction or a decimal as a percent

A decimal or a fraction can be written as a percent by multiplying by 100%.

Write $\frac{3}{8}$ as a percent.

$\frac{3}{8} = \frac{3}{8} \times 100\% = \frac{3}{8} \times \frac{100}{1}\% = \frac{300}{8}\% = 37\frac{1}{2}\%$ or 37.5%

Write 0.37 as a percent.

$0.37 = 0.37 \times 100\% = 37\%$

> Move the decimal point two places to the right. Then write the percent sign.

Example 7 Write 0.015 as a percent.
Solution $0.015 = 0.015 \times 100\%$
$= 1.5\%$

Example 8 Write 0.048 as a percent.
Your solution

Example 9 Write 2.15 as a percent.
Solution $2.15 = 2.15 \times 100\% = 215\%$

Example 10 Write 3.67 as a percent.
Your solution

Example 11 Write $0.33\frac{1}{3}$ as a percent.
Solution $0.33\frac{1}{3} = 0.33\frac{1}{3} \times 100\%$
$= 33\frac{1}{3}\%$

Example 12 Write $0.62\frac{1}{2}$ as a percent.
Your solution

Example 13 Write $\frac{2}{3}$ as a percent. Write the remainder in fractional form.
Solution $\frac{2}{3} = \frac{2}{3} \times 100\% = \frac{200}{3}\%$
$= 66\frac{2}{3}\%$

Example 14 Write $\frac{5}{6}$ as a percent. Write the remainder in fractional form.
Your solution

Example 15 Write $2\frac{2}{7}$ as a percent. Round to the nearest tenth.
Solution $2\frac{2}{7} = \frac{16}{7} = \frac{16}{7} \times 100\%$
$= \frac{1600}{7}\% \approx 228.6\%$

Example 16 Write $1\frac{4}{9}$ as a percent. Round to the nearest tenth.
Your solution

Solutions on p. A21

5.1 EXERCISES

▶ **Objective A**

Write as a fraction and as a decimal.

1. 25%
2. 40%
3. 130%
4. 150%
5. 100%
6. 87%
7. 73%
8. 45%
9. 383%
10. 425%
11. 70%
12. 55%
13. 88%
14. 64%
15. 32%
16. 18%

Write as a fraction.

17. $66\frac{2}{3}\%$
18. $12\frac{1}{2}\%$
19. $83\frac{1}{3}\%$
20. $3\frac{1}{8}\%$
21. $11\frac{1}{9}\%$
22. $\frac{3}{8}\%$
23. $45\frac{5}{11}\%$
24. $15\frac{3}{8}\%$
25. $4\frac{2}{7}\%$
26. $5\frac{3}{4}\%$
27. $6\frac{2}{3}\%$
28. $8\frac{2}{3}\%$

Write as a decimal.

29. 6.5%
30. 12.3%
31. 0.55%
32. 2%
33. 8.25%
34. 5.05%
35. 6.75%
36. 3.08%
37. 0.45%
38. 6.4%
39. 80.4%
40. 16.7%

Objective B

Write as a percent.

41. 0.16 42. 0.73 43. 0.05 44. 0.13 45. 0.01 46. 0.95

47. 0.70 48. 1.07 49. 1.24 50. 2.07 51. 0.004 52. 0.37

53. 0.006 54. 1.012 55. 3.106 56. 0.12

Write as a percent. Round to the nearest tenth.

57. $\frac{27}{50}$ 58. $\frac{37}{100}$ 59. $\frac{1}{3}$ 60. $\frac{2}{5}$

61. $\frac{5}{8}$ 62. $\frac{1}{8}$ 63. $\frac{1}{6}$ 64. $1\frac{1}{2}$

65. $\frac{7}{40}$ 66. $1\frac{2}{3}$ 67. $1\frac{7}{9}$ 68. $\frac{7}{8}$

Write as a percent. Write the remainder in fractional form.

69. $\frac{15}{50}$ 70. $\frac{12}{25}$ 71. $\frac{7}{30}$ 72. $\frac{1}{3}$

73. $\frac{4}{7}$ 74. $\frac{1}{12}$ 75. $\frac{3}{11}$ 76. $\frac{2}{9}$

77. $2\frac{3}{8}$ 78. $1\frac{2}{3}$ 79. $2\frac{1}{6}$ 80. $\frac{7}{8}$

SECTION 5.2 Percent Equations: Part I

Objective A To find the amount when the percent and base are given

A real estate broker receives a payment that is 4% of an $85,000 sale. To find the amount the broker receives requires answering the question "4% of $85,000 is what?"

This sentence can be written using mathematical symbols and then solved for the unknown number.

4% of $85,000 is what?

percent	×	base	=	amount
4%		$85,000		n

of is written as × (times)
is is written as = (equals)
what is written as n (the unknown number)

$0.04 \times \$85,000 = n$
$\$3400 = n$

Notice that the percent is written as a decimal.

The broker receives a payment of $3400.

The solution was found by solving the basic percent equation for amount.

The Basic Percent Equation

| Percent | × | base | = | amount |

In most cases, the percent is written as a decimal before the basic percent equation is solved. However, some percents are more easily written as a fraction than as a decimal. For example,

$33\frac{1}{3}\% = \frac{1}{3}$ $66\frac{2}{3}\% = \frac{2}{3}$ $16\frac{2}{3}\% = \frac{1}{6}$ $83\frac{1}{3}\% = \frac{5}{6}$

Example 1 Find 5.7% of 160.

Solution $n = 0.057 \times 160$
$n = 9.12$

Note the words "what is" are missing from the problem but are implied by the word "Find."

Example 2 Find 6.3% of 150.

Your solution

Example 3 What is $33\frac{1}{3}\%$ of 90?

Solution $n = \frac{1}{3} \times 90$
$n = 30$

Example 4 What is $16\frac{2}{3}\%$ of 66?

Your solution

Solutions on p. A21

Objective B

The solution to percent problems requires identifying the three elements of the basic percent equation. Recall that these three parts are percent, base, and amount. Usually the base will follow the phrase "percent of."

Example 5

The harvest from a small orange grove was 560 crates of oranges. 5% of the 560 crates were damaged during shipment. How many crates were damaged?

Strategy

To find how many crates were damaged, write and solve a basic percent equation using n to represent the number of crates damaged (amount). The percent is 5% and the base is 560.

Solution $5\% \times 560 = n$
$0.05 \times 560 = n$
$28 = n$

28 crates were damaged during shipment.

Example 7

A quality control inspector found that 1.2% of 2500 telephones inspected were defective. How many telephones were not defective?

Strategy

To find the number of nondefective phones:

- Find the number of defective phones. Write and solve a basic percent equation using n to represent the number of defective phones (amount). The percent is 1.2% and the base is 2500.
- Subtract the number of defective phones from the number of phones inspected (2500).

Solution $1.2\% \times 2500 = n$
$0.012 \times 2500 = n$
$30 = n$ defective phones

$2500 - 30 = 2470$

2470 telephones were not defective.

Example 6

A person buys a television that costs $450 and pays a sales tax that is 6% of the cost. What is the sales tax on the television?

Your strategy

Your solution

Example 8

An electrician's hourly wage is $13.50 before an 8% raise. What is the new hourly wage?

Your strategy

Your solution

Solutions on p. A21

5.2 EXERCISES

▶ Objective A

Solve.

1. 8% of 100 is what?

2. 16% of 50 is what?

3. 27% of 40 is what?

4. 52% of 95 is what?

5. 0.05% of 150 is what?

6. 0.075% of 625 is what?

7. 125% of 64 is what?

8. 210% of 12 is what?

9. Find 10.7% of 485.

10. Find 12.8% of 625.

11. What is 0.25% of 3000?

12. What is 0.06% of 250?

13. 80% of 16.25 is what?

14. 26% of 19.5 is what?

15. What is $1\frac{1}{2}$% of 250?

16. What is $5\frac{3}{4}$% of 65?

17. $16\frac{2}{3}$% of 120 is what?

18. $83\frac{1}{3}$% of 246 is what?

19. What is $33\frac{1}{3}$% of 630?

20. What is $66\frac{2}{3}$% of 891?

21. Which is larger: 5% of 95, or 75% of 6?

22. Which is larger: 82% of 16, or 20% of 65?

23. Which is larger: 22% of 120, or 84% of 32?

24. Which is larger: 112% of 5, or 0.45% of 800?

25. Which is smaller: 15% of 80, or 95% of 15?

26. Which is smaller: 2% of 1500, or 72% of 40?

27. Find 31.294% of 82,460.

28. Find 123.94% of 275,976.

▶ Objective B   *Application Problems*

29. A salesclerk receives a salary of $2240 per month, and 18% of this amount is deducted for income tax. Find the amount deducted for income tax.

30. In a city election for mayor, the successful candidate received 53% of the 385,000 votes cast. How many votes did the successful candidate receive?

31. An antique shop owner expects to receive $16\frac{2}{3}$% of the shop's sales as profit. What is the profit in a month when the total sales are $24,000?

32. A student survey in a community college found that $33\frac{1}{3}$% of its 7500 beginning students did not complete their first year of college. How many of the beginning students did not complete the first year of college?

33. A recreational vehicle dealer offers a 7% rebate on some RV models. What rebate would a buyer receive on an RV that cost $24,000?

34. A farmer is given an income tax credit of 10% of the cost of some farm machinery. What tax credit would the farmer receive on farm equipment that cost $85,000?

35. You purchase a car for $9500 and must pay a sales tax of 6% of the cost.
    a. What is the sales tax?
    b. What is the total cost of the car including sales tax?

36. During the packaging process for oranges, spoiled oranges are discarded by an inspector. In 1 day an inspector found that 4.8% of the 20,000 pounds of oranges inspected were spoiled.
    a. How many pounds of oranges were spoiled?
    b. How many pounds of oranges were not spoiled?

37. An amusement park has 550 employees and must hire an additional 22% for the vacation season. What is the total number of employees needed for the vacation season?

38. An entertainment system was purchased for $1350. A down payment of 15% was required. Find the amount remaining to be paid after the down payment had been paid.

39. An office building has an appraised value of $2,862,400. The real estate taxes are 1.75% of the appraised value of the building. Find the real estate taxes.

40. A company pays out 35% of its profits in dividends. Find the dividends paid out to investors when the company had a profit of $22,540,000.

SECTION 5.3 Percent Equations: Part II

Objective A — To find the percent when the base and amount are given

A recent promotional game at a grocery store listed the probability of winning a prize as "1 chance in 2." A percent can be used to describe the chance of winning. This requires answering the question, "What percent of 2 is 1?"

The chance of winning can be found by solving the basic percent equation for *percent*.

What percent of 2 is 1

$$\text{percent} \times \text{base} = \text{amount}$$
$$n \times 2 = 1$$
$$n \times 2 = 1$$
$$n = 1 \div 2$$
$$n = 0.5$$
$$n = 50\%$$

The solution must be written as a percent to answer the question.

There is a 50% chance of winning a prize.

Example 1 What percent of 40 is 30?

Solution
$$n \times 40 = 30$$
$$n = 30 \div 40$$
$$n = 0.75$$
$$n = 75\%$$

Example 2 What percent of 32 is 16?

Your solution

Example 3 What percent of 12 is 27?

Solution
$$n \times 12 = 27$$
$$n = 27 \div 12$$
$$n = 2.25$$
$$n = 225\%$$

Example 4 What percent of 15 is 48?

Your solution

Example 5 25 is what percent of 75?

Solution
$$25 = n \times 75$$
$$25 \div 75 = n$$
$$0.33\tfrac{1}{3} = n$$
$$33\tfrac{1}{3}\% = n$$

Example 6 30 is what percent of 45?

Your solution

Solutions on p. A22

Objective B To solve application problems

To solve percent problems, remember that it is necessary to identify the percent, base, and amount. Usually the base will follow the phrase "percent of."

Example 7
A family has an income of $1500 each month and makes a car payment of $180 a month. What percent of the income is the car payment?

Strategy
To find what percent of the income the car payment is, write and solve the basic percent equation using n to represent the unknown percent. The base is $1500 and the amount is $180.

Solution $n \times \$1500 = \$180$
$n = \$180 \div \$1500$
$n = 0.12 = 12\%$

The car payment is 12% of the monthly income.

Example 8
A housepainter had an income of $20,000 and paid $3000 in income tax. What percent of the income is the income tax?

Your strategy

Your solution

Example 9
A nurse missed 25 questions out of 200 on a state license exam. What percent of the questions did the nurse answer correctly?

Strategy
To find what percent of the questions the nurse answered correctly:
- Find the number of questions the nurse answered correctly (200 − 25).
- Write and solve a basic percent equation using n to represent the unknown percent. The base is 200 and the amount is the number of questions answered correctly.

Solution 200 − 25 = 175 questions answered correctly

$n \times 200 = 175$
$n = 175 \div 200$
$n = 0.875 = 87.5\%$

The nurse answered 87.5% of the questions correctly.

Example 10
A survey of 1000 people showed that 667 people favored a candidate for governor of the state. What percent of the people surveyed did not favor the candidate?

Your strategy

Your solution

Solutions on p. A22

5.3 EXERCISES

▶ Objective A

Solve.

1. What percent of 75 is 24?
2. What percent of 80 is 20?
3. 15 is what percent of 90?
4. 24 is what percent of 60?
5. What percent of 12 is 24?
6. What percent of 6 is 9?
7. What percent of 16 is 6?
8. What percent of 24 is 18?
9. 18 is what percent of 100?
10. 54 is what percent of 100?
11. 5 is what percent of 2000?
12. 8 is what percent of 2500?
13. What percent of 6 is 1.2?
14. What percent of 2.4 is 0.6?
15. 16.4 is what percent of 4.1?
16. 5.3 is what percent of 50?
17. 1 is what percent of 40?
18. 0.3 is what percent of 20?
19. What percent of 48 is 18?
20. What percent of 11 is 88?
21. What percent of 2800 is 7?
22. What percent of 400 is 12?
23. 4.2 is what percent of 175?
24. 41.79 is what percent of 99.5?
25. What percent of 86.5 is 8.304?
26. What percent of 1282.5 is 2.565?
27. 632,000 is what percent of 1,469,300?
28. 16.235 is what percent of 19.6375?

Objective B Application Problems

29. A company spends $4500 of its $90,000 budget for advertising. What percent of the budget is spent for advertising?

30. Last month a thrift store had an income of $2812.50. The thrift store pays $900 a month for rent. What percent of last month's income was spent for rent?

31. A car is sold for $8900. The buyer of the car pays a down payment of $1780. What percent of the selling price is the down payment?

32. A utility company's stock is trading at 24\frac{1}{2}$ and pays a dividend of $1.96. What percent of the stock price is the dividend?

33. The value of your home in 1984 was $78,500. The value of your home in 1990 was $141,300. What percent of the 1984 value is the 1990 value?

34. A television survey of 4000 families found that 2500 liked a new TV show. What percent of the families surveyed liked the new TV show?

35. A house in a fast-growing city in the southwest was bought for $124,000. A year later the house was sold for $155,000.
    a. Find the increase in price.
    b. What percent of the original price is the increase?

36. A 3.5-pound beef roast contains 0.7 pounds of fat.
    a. What percent of the beef roast is fat?
    b. At the same rate, how many pounds of fat would be in a side of beef weighing 550 pounds?

37. To receive a license to sell insurance, an insurance account executive must answer 70% of the 250 questions on a test correctly. An account executive answered 177 questions correctly. Was this amount enough to pass the test?

38. A test of the breaking strength of concrete slabs for freeway construction found that 3 of the 200 tested did not meet safety requirements. What percent of the slabs tested did meet safety requirements?

39. The surface area of the earth, including land and water, is approximately 197,470,000 square miles. The land area of the earth is estimated as 57,266,300 square miles. What percent of the total area of the earth is the land area?

40. A 50.5-acre piece of raw land is appraised at a value of $5,900,000. The land is bought for $4,950,000. What percent of the appraised value is the purchase price? (Round to the nearest tenth of a percent.)

SECTION 5.4 Percent Equations: Part III

Objective A To find the base when the percent and amount are given

Each year an investor receives a payment that equals 12% of the value of an investment. This year that payment amounted to $480. To find the value of the investment this year, the investor must find "12% of what value is $480?"

The value of the investment can be found by solving the basic percent equation for the base.

12% of what is $480

percent 12% × base n = amount $480

$$0.12 \times n = \$480$$
$$n = \$480 \div 0.12$$
$$n = \$4000$$

This year the investment is worth $4000.

Example 1 18% of what is 900?

Solution $0.18 \times n = 900$
$n = 900 \div 0.18$
$n = 5000$

Example 2 86% of what is 215?

Your solution

Example 3 30 is 1.5% of what?

Solution $30 = 0.015 \times n$
$30 \div 0.015 = n$
$2000 = n$

Example 4 15 is 2.5% of what?

Your solution

Example 5 $33\frac{1}{3}\%$ of what is 7?

Solution $\frac{1}{3} \times n = 7$ Note the
$n = 7 \div \frac{1}{3}$ percent is written as
$n = 21$ a fraction.

Example 6 $16\frac{2}{3}\%$ of what is 5?

Your solution

Solutions on p. A22

Objective B To solve application problems

To solve percent problems, it is necessary to identify the percent, base, and amount. Usually the base will follow the phrase "percent of."

Example 7

A business office bought a used copy machine for $450, which was 75% of the original cost. What was the original cost of the copier?

Strategy

To find the original cost of the copier, write and solve the basic percent equation using n to represent the original cost (base). The percent is 75% and the amount is $450.

Solution

$75\% \times n = \$450$
$0.75 \times n = \$450$
$n = \$450 \div 0.75$
$n = \$600$

The original cost of the copier was $600.

Example 8

A used car has a value of $3876, which is 51% of the car's original value. What was the car's original value?

Your strategy

Your solution

Example 9

A carpenter's wage this year is $19.80 per hour, which is 110% of last year's wage. What was the increase in the hourly wage over last year?

Strategy

To find the increase in the hourly wage over last year:

- Find last year's wage. Write and solve the basic percent equation using n to represent last year's wage (base). The percent is 110% and the amount is $19.80.
- Subtract last year's wage from this year's wage ($19.80).

Solution

$110\% \times n = \$19.80$
$1.10 \times n = \$19.80$
$n = \$19.80 \div 1.10$
$n = \$18.00$ last year's wage

$\$19.80 - \$18.00 = \$1.80$

The increase in the hourly wage is $1.80.

Example 10

A sporting goods store has a tennis racket on sale for $44.80, which is 80% of the original price. What is the difference between the original price and the sale price?

Your strategy

Your solution

Solutions on pp. A22–A23

5.4 EXERCISES

▶ Objective A

Solve.

1. 12% of what is 9?

2. 38% of what is 171?

3. 8 is 16% of what?

4. 54 is 90% of what?

5. 10 is 10% of what?

6. 37 is 37% of what?

7. 30% of what is 25.5?

8. 25% of what is 21.5?

9. 2.5% of what is 30?

10. 10.4% of what is 52?

11. 125% of what is 24?

12. 180% of what is 21.6?

13. 18 is 240% of what?

14. 24 is 320% of what?

15. 4.8 is 15% of what?

16. 87.5 is 50% of what?

17. 25.6 is 12.8% of what?

18. 45.014 is 63.4% of what?

19. 0.7% of what is 0.56?

20. 0.25% of what is 1?

21. 30% of what is 2.7?

22. 78% of what is 3.9?

23. 84 is $16\frac{2}{3}$% of what?

24. 120 is $33\frac{1}{3}$% of what?

25. $66\frac{2}{3}$% of what is 72?

26. $83\frac{1}{3}$% of what is 13.5?

27. 6.59% of what is 469.35?

28. 182.3% of what is 46,253?

Objective B Application Problems

29. A mechanic estimates that the brakes of an RV still have 6000 miles of wear. This amount is 12% of the estimated safe-life use of the brakes. What is the estimated life of the brakes?

30. A used mobile home was purchased for $18,000. This amount was 64% of the new mobile home cost. What was the cost of a new mobile home?

31. A city has a population of 42,000. This amount is 75% of what the population was 5 years ago. What was the city's population 5 years ago?

32. A ski resort had a snowfall of 198 inches during a year. This amount is 110% of the previous year's snowfall. What was the previous year's snowfall?

33. A store advertised a scientific calculator for $55.80. This amount was 120% of the cost at a competing store. What is the price at the competitor's store?

34. A salesperson receives a commission of $820 for selling a car. This amount is 5% of the selling price of the car. What is the selling price of the car?

35. During a quality control test, a manufacturer of computer boards found that 24 boards were defective. This amount was 0.8% of the computer boards tested.
    a. How many computer boards were tested?
    b. How many computer boards tested were not defective?

36. Of the calls a directory assistance operator received, 441 were requests for telephone numbers listed in the current directory. This amount was 98% of the calls for assistance that the operator received.
    a. How many calls did the operator receive?
    b. How many telephone numbers requested were not listed in the current directory?

37. An orange grower produced 22,680 boxes of oranges this year. This amount was 108% of last year's harvest. What was the increase in the number of boxes of oranges over last year?

38. A steelworker's wage this year is $16.80 an hour. This amount is 112% of last year's wage. What was the increase in the hourly wage over last year?

39. A corporation earned $6,512,400 during the year. This amount was 120% of last year's earnings. Find the increase in the corporation's earnings for the year.

40. Defects were found in 5400 diodes. This amount was 0.12% of the total number of diodes produced in 1 month. Find the total number of diodes produced in that month.

SECTION 5.5 Percent Problems: Proportion Method

Objective A — To solve percent problems using proportions

The percent problems in Sections 5.2 through 5.4 were solved using the basic percent equation. Those problems can also be solved using proportions.

The proportion method is based on writing two ratios. One ratio is the percent ratio, written as $\frac{percent}{100}$. The second ratio is the amount-to-base ratio, written as $\frac{amount}{base}$. These two ratios form the proportion

$$\frac{percent}{100} = \frac{amount}{base}$$

To use the proportion method, first identify the percent, the amount, and the base (the base usually follows the phrase "percent of").

What is 23% of 45?

$$\frac{23}{100} = \frac{n}{45}$$
$23 \times 45 = 100 \times n$
$1035 = 100 \times n$
$1035 \div 100 = n$
$10.35 = n$

What percent of 25 is 4?

$$\frac{n}{100} = \frac{4}{25}$$
$n \times 25 = 100 \times 4$
$n \times 25 = 400$
$n = 400 \div 25$
$n = 16$

12 is 60% of what number?

$$\frac{60}{100} = \frac{12}{n}$$
$60 \times n = 100 \times 12$
$60 \times n = 1200$
$n = 1200 \div 60$
$n = 20$

Example 1 15% of what is 7? Round to the nearest hundredth.

Solution
$$\frac{15}{100} = \frac{7}{n}$$
$15 \times n = 100 \times 7$
$15 \times n = 700$
$n = 700 \div 15$
$n \approx 46.67$

Example 2 26% of what is 22? Round to the nearest hundredth.

Your solution

Example 3 30% of 63 is what?

Solution
$$\frac{30}{100} = \frac{n}{63}$$
$30 \times 63 = n \times 100$
$1890 = n \times 100$
$1890 \div 100 = n$
$18.90 = n$

Example 4 16% of 132 is what?

Your solution

Solutions on p. A23

Objective B To solve application problems

Example 5

An antique dealer found that 86% of the 250 items that were sold during 1 month sold for under $1000. How many items sold for under $1000?

Strategy

To find the number of items that sold for under $1000, write and solve a proportion using n to represent the number of items sold (amount) for less than $1000. The percent is 86% and the base is 250.

Solution
$$\frac{86}{100} = \frac{n}{250}$$
$$86 \times 250 = 100 \times n$$
$$21{,}500 = 100 \times n$$
$$21{,}500 \div 100 = n$$
$$215 = n$$

215 items sold for under $1000.

Example 6

Last year it snowed 64% of the 150 days of the ski season at a resort. How many days did it snow?

Your strategy

Your solution

Example 7

In a test of the strength of nylon rope, 5 pieces of the 25 pieces tested did not meet the test standards. What percent of the nylon ropes tested did meet the standards?

Strategy

To find the percent of ropes tested that met the standards:

- Find the number of ropes that met the test standards (25 − 5).
- Write and solve a proportion using n to represent the percent of ropes that met the test standards. The base is 25. The amount is the number of ropes that met the standards.

Solution 25 − 5 = 20 ropes met test standards
$$\frac{n}{100} = \frac{20}{25}$$
$$n \times 25 = 100 \times 20$$
$$n \times 25 = 2000$$
$$n = 2000 \div 25$$
$$n = 80$$

80% of the ropes tested did meet the test standards.

Example 8

Five ballpoint pens in a box of 200 were found to be defective. What percent of the pens were not defective?

Your strategy

Your solution

Solutions on p. A23

5.5 EXERCISES

▶ **Objective A**

Solve.

1. 26% of 250 is what?
2. What is 18% of 150?
3. 37 is what percent of 148?
4. What percent of 150 is 33?
5. 68% of what is 51?
6. 126 is 84% of what?
7. What percent of 344 is 43?
8. 750 is what percent of 50?
9. 82 is 20.5% of what?
10. 2.4% of what is 21?
11. What is 6.5% of 300?
12. 96% of 75 is what?
13. 7.4 is what percent of 50?
14. What percent of 1500 is 693?
15. 50.5% of 124 is what?
16. What is 87.4% of 255?
17. 120% of what is 6?
18. 14 is 175% of what?
19. What is 250% of 18?
20. 325% of 4.4 is what?
21. 33 is 220% of what?
22. 160% of what is 40?
23. 87 is what percent of 29?
24. What percent of 38 is 95?
25. 0.25% of what is 8?
26. 14 is 0.5% of what?
27. What percent of 250 is 6?
28. 15 is what percent of 5000?
29. Find 1.237% of 996,254.
30. 18,572 is what percent of 1,325,500?

Objective B Application Problems

31. A charity organization spent $2940 for administrative expenses. This amount is 12% of the money it collected. What is the total amount the charity organization collected?

32. A manufacturer of an anti-inflammatory drug claims that the drug will be effective for 6 hours. An independent testing service determined that the drug was effective only 80% of the length of time claimed by the manufacturer. Find the length of time the drug will be effective as determined by the testing service.

33. A calculator can be purchased for $28.50. This amount is 40% of the cost of the calculator 8 years ago. What was the cost of the calculator 8 years ago?

34. A produce market increased the price of lettuce to $.69 per pound, which is 115% of the original price. What was the original price?

35. A fire department received 24 false alarms out of a total of 200 alarms received. What percent of the alarms received were false alarms?

36. A typist made errors on 5 words on a typing test. The total length of the test was 250 words. What percent of the words were typed correctly?

37. It cost a mechanic $174 to repair a car. Of this cost, 55% was for labor and 45% was for parts.
    a. What was the labor cost?
    b. What was the total cost for parts?

38. In the last 500 hours of a computer operation, the computer malfunctioned 0.8% of the time.
    a. How many hours did the computer malfunction?
    b. How many hours did the computer operate properly?

39. In a test of a new antihistamine with 800 volunteers with allergy problems, 60 did not show any improvement. What percent of the volunteers did show improvement?

40. A monthly state income tax is 11% of the amount over $1700. What state income tax does a person pay on a salary of $2600?

41. A baseball player with a salary of $1,650,000 is offered a new contract that contains a 11.25% pay cut. What is the amount of money offered in the new contract?

42. You bought an automobile 1 year ago. Since then it has depreciated $932.09, which is 11.23% of the price you paid for it. How much did you pay for the automobile?

Calculators and Computers

The Percent Key on a Calculator

Finding the discount and sales price of an item that is on sale is one of the many uses of the percent key on a calculator.

A percent may be entered on a calculator in either of the following two ways:

1. The first way to enter a percent on a calculator is to enter it as a decimal. Remember, to write a percent as a decimal, remove the percent sign and multiply by 0.01. For example, to multiply a number by 12%, use the decimal 0.12.
2. The second way to enter a percent on a calculator is to use the $\boxed{\%}$ key.

The following two examples demonstrate the procedure for using this key.

Example 1 A car stereo that normally sells for $98.95 is on sale for 15% off the regular price. Find the discount and the sales price.

Enter	Display	Comments
98.95	98.95	Enter the regular price.
$\boxed{\times}$	98.95	Multiply the regular price by the percent.
15	15	Enter the percent discount.
$\boxed{\%}$	14.8425	The discount is $14.84 when rounded to the nearest cent.
$\boxed{-}$	84.1075	This step will work only if your calculator is a scientific calculator and programmed to use the Order of Operations Agreement.
		The sales price is $84.11.

Example 2 A new tire for a car costs $45.75. If the sales tax rate is 6%, find the sales tax and the total cost.

Enter	Display	Comments
45.75	45.75	Enter the cost.
$\boxed{\times}$	45.75	Multiply the cost by the sales tax rate.
6	6.	Enter the sales tax rate.
$\boxed{\%}$	2.745	The sales tax is $2.75.
$\boxed{+}$	48.495	Add the sales tax to the cost to get the total cost.
		The total cost is $48.50.

Chapter Summary

Key Words *Percent* means "parts of 100."

Essential Rules *To Write a Percent as a Fraction*

To write a percent as a fraction, remove the percent sign and multiply by $\frac{1}{100}$.

To Write a Percent as a Decimal

To write a percent as a decimal, remove the percent sign and multiply by 0.01.

To Write a Decimal as a Percent

To write a decimal as a percent, multiply by 100%.

To Write a Fraction as a Percent

To write a fraction as a percent, multiply by 100%.

Chapter Review

SECTION 5.1

1. Write 12% as a fraction.

2. Write $16\frac{2}{3}$% as a fraction.

3. Write 42% as a decimal.

4. Write 7.6% as a decimal.

5. Write 0.38 as a percent.

6. Write $1\frac{1}{2}$ as a percent.

SECTION 5.2

7. What is 30% of 200?

8. What is 7.5% of 72?

9. Find 22% of 88.

10. Find 125% of 62.

11. A company uses 7.5% of its $60,000 expense budget for TV advertising. How much of the company's expense budget was spent for TV advertising?

12. A customer purchases a video camera for $980 and must pay a sales tax of 6.25% of the cost. What is the total cost of the video camera?

SECTION 5.3

13. What percent of 20 is 30?

14. 16 is what percent of 80?

15. What percent of 30 is 2.2? Round to the nearest tenth of a percent.

16. What percent of 15 is 92? Round to the nearest tenth of a percent.

17. A company's stock is trading at $65 and paying a dividend of $2.99. What percent of the stock price is the dividend?

18. A house in southern California had a value of $125,000. In 2 years the house was valued at $205,000. Find the percent increase in value during the 2 years.

SECTION 5.4

19. 20% of what is 15?

20. 78% of what is 8.5? Round to the nearest tenth.

21. $66\frac{2}{3}\%$ of what is 105?

22. $16\frac{2}{3}\%$ of what is 84?

23. A city's population this year is 157,500, which is 225% of what the city's population was 10 years ago. What was the city's population 10 years ago?

24. A baseball player has a batting average of 0.315. This average is an increase of 125% over the previous year's batting average. Find the previous year's batting average.

SECTION 5.5

25. What is 62% of 320?

26. What percent of 25 is 40?

27. A computer system can be purchased for $1800. This price is 60% of what the computer cost 4 years ago. What was the cost of the computer 4 years ago?

28. A student misses 9 out of 60 questions on a history exam. What percent of the questions did the student answer correctly?

Chapter Test

1. Write 97.3% as a decimal.

2. Write $16\frac{2}{3}\%$ as a fraction.

3. Write 0.3 as a percent.

4. Write 1.63 as a percent.

5. Write $\frac{3}{2}$ as a percent.

6. Write $\frac{2}{3}$ as a percent.

7. What is 77% of 65? Round to the nearest hundredth.

8. 47.2% of 130 is what? Round to the nearest hundredth.

9. Which is larger:
   7% of 120, or 76% of 13?

10. Which is smaller:
    13% of 200, or 212% of 12?

11. A fast-food company uses 6% of its $75,000 budget for advertising. What amount of the budget is spent for advertising?

12. During the packaging process for vegetables, spoiled vegetables are discarded by an inspector. In 1 day an inspector found that 6.4% of the 1250 pounds of vegetables were spoiled. How many pounds of vegetables were not spoiled?

13. What percent of 120 is 30?

14. 26 is what percent of 12? Round to the nearest hundredth.

15. A department store has 125 permanent employees and must hire an additional 20 temporary employees for the holiday season. What percent of the permanent employees is hired as temporary employees for the holiday season?

16. A student missed 7 out of 80 questions on a math exam. What percent of the questions did the student answer correctly? (Round to the nearest tenth of a percent)

17. 12 is 15% of what?

18. 42.5 is 150% of what? Round to the nearest tenth.

19. A manufacturer of transistors found 384 defective transistors during a quality control study. This amount was 1.2% of the transistors tested. Find the number of transistors tested.

20. A new house was bought for $95,000; 5 years later the house sold for $152,000. The increase is what percent of the original price?

21. 123 is 86% of what number? Round to the nearest tenth.

22. What percent of 12 is 120?

23. A secretary receives a wage of $9.52 per hour. This amount is 112% of last year's salary. What is the dollar increase in the hourly wage over last year?

24. A city has a population of 71,500; 10 years ago the population was 32,500. The population now is what percent of what the population was 10 years ago?

25. The annual license fee on a car is 1.4% of the value of the car. If the license fee during a year was $91.00, what is the value of the car?

Cumulative Review

1. Simplify $18 \div (7 - 4)^2 + 2$.

2. Find the LCM of 16, 24, and 30.

3. Add: $2\frac{1}{3} + 3\frac{1}{2} + 4\frac{5}{8}$

4. Subtract: $27\frac{5}{12} - 14\frac{9}{16}$

5. Multiply: $7\frac{1}{3} \times 1\frac{5}{7}$

6. Divide: $\frac{14}{27} \div 1\frac{7}{9}$

7. Simplify: $\left(\frac{3}{4}\right)^3 \cdot \left(\frac{8}{9}\right)^2$

8. Simplify: $\left(\frac{2}{3}\right)^2 - \left(\frac{3}{8} - \frac{1}{3}\right) \div \frac{1}{2}$

9. Round 3.07973 to the nearest hundredth.

10. Subtract: 3.0902
               −1.9706

11. Divide: $0.032 \overline{)1.097}$
    Round to the nearest ten-thousandth.

12. Convert $3\frac{5}{8}$ to a decimal.

13. Convert 1.75 to a fraction.

14. Place the correct symbol, < or >, between the two numbers.
    $\frac{3}{8}$ 0.87

15. Solve the proportion $\frac{3}{8} = \frac{20}{n}$. Round to the nearest tenth.

16. Write "$76.80 earned in 8 hours" as a unit rate.

17. Write $18\tfrac{1}{3}\%$ as a fraction.

18. Write $\tfrac{5}{6}$ as a percent.

19. 16.3% of 120 is what? Round to the nearest hundredth.

20. 24 is what percent of 18?

21. 12.4 is 125% of what?

22. What percent of 35 is 120? Round to the nearest tenth.

23. A mechanic has an income of $740 per week. One-fifth of the income is deducted for income tax payments. Find the mechanic's take-home pay.

24. A student buys a car for $4321, with a down payment of $1000. The balance is paid in 36 equal monthly payments. Find the monthly payment.

25. Gasoline tax is $0.19 a gallon. Find the number of gallons of gasoline used during a month in which $79.80 was paid in taxes.

26. The real estate tax on a $72,000 home is $1440. At the same rate, find the real estate tax on a home valued at $150,000.

27. A customer purchases a stereo set for $490 and pays $29.40 in sales tax. What percent of the purchase price is the sales tax?

28. A survey of 300 people showed that 165 people favored a candidate for mayor. What percent of the people surveyed did not favor the candidate?

29. The value of a home in the northern part of the United States was $62,000 in 1985. The same home in 1990 had a value of $155,000. What percent of the 1985 value is the 1990 value?

30. The Environmental Protection Agency found that 990 out of 5500 children tested had levels of lead in their blood exceeding federal guidelines. What percent of the children tested had levels of lead in the blood that exceeded federal standards?

6 Applications for Business and Consumers: A Calculator Approach

OBJECTIVES
- To find unit cost
- To find the most economical purchase
- To find total cost
- To find percent increase
- To apply percent increase to business—markup
- To find percent decrease
- To apply percent decrease to business—discount
- To calculate simple interest
- To calculate compound interest
- To calculate the initial expenses of buying a home
- To calculate ongoing expenses of owning a home
- To calculate the initial expenses of buying a car
- To calculate ongoing expenses of owning a car
- To calculate commissions, total hourly wages, and salaries
- To calculate checkbook balances
- To balance a checkbook

A Penny a Day

A fictitious job offer in a newspaper claimed that it would pay a salary of 1¢ on the first of the month and 2¢ on the second of the month. Next month, the salary would be 4¢ on the first and 8¢ on the second of the month. The next month and each succeeding month for 12 months the procedure of doubling the previous payment would be continued the same way. Do you think you would want the job under these conditions?

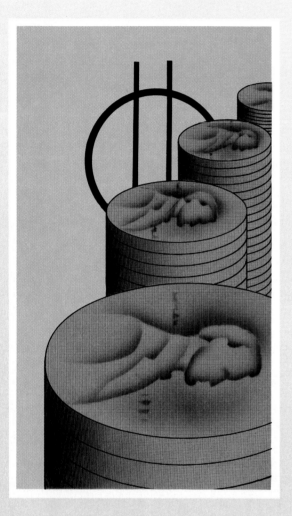

This problem is an example of compounding interest, which is similar to the type of interest that is earned on bank or savings deposits. The big difference is that banks and savings and loans do not compound the interest as quickly as in the given example.

The table below shows the amount of salary you would earn for each month and then gives the total annual salary.

January	$0.01 + $0.02 = $0.03
February	$0.04 + $0.08 = $0.12
March	$0.16 + $0.32 = $0.48
April	$0.64 + $1.28 = $1.92
May	$2.56 + $5.12 = $7.68
June	$10.24 + $20.48 = $30.72
July	$40.96 + $81.92 = $122.88
August	$163.84 + $327.68 = $491.52
September	$655.36 + $1310.72 = $1966.08
October	$2621.44 + $5242.88 = $7864.32
November	$10,485.76 + $20,971.52 = $31,457.28
December	$41,943.04 + $83,886.08 = $125,829.12

Total Annual Salary = $167,772.15

Not a bad annual salary!

Section 6.1 / Applications to Purchasing

SECTION 6.1 Applications to Purchasing

Objective A To find unit cost

Frequently stores advertise items for purchase as

3 shirts for $36 or

5 pounds of potatoes for 90¢.

The **unit cost** is the cost of 1 shirt or 1 pound of potatoes. To find the unit cost, divide the total cost by the number of units.

3 shirts for $36	5 pounds of potatoes for 90¢
36 ÷ 3 = 12	90 ÷ 5 = 18
$12 is the cost of 1 shirt.	18¢ is the cost of 1 pound of potatoes.
Unit cost: $12 per shirt	Unit cost: 18¢ per pound

Example 1 Find the unit cost. Round to the nearest tenth of a cent.
a) 10 gallons of gasoline for $15.49
b) 3 yards of material for $10.

Strategy To find the unit cost, divide the total cost by the number of units.

Solution
a) 15.49 ÷ 10 = 1.549
   $1.549 per gallon
b) 10 ÷ 3 ≈ 3.3333
   $3.333 per yard

Example 2 Find the unit cost. Round to the nearest tenth of a cent.
a) 5 quarts of oil for $6.25
b) 4 ears of corn for 85¢

Your strategy

Your solution

Solution on p. A24

Objective B To find the most economical purchase

Comparison shoppers often find the most economical buy by comparing unit costs.

One store sells 6 cans of cola for $2.04 and another store sells 24 cans of the same brand for $7.92. To find the better buy, compare the unit costs.

2.04 ÷ 6 = 0.34	7.92 ÷ 24 = 0.33
Unit cost: $0.34 per can	Unit cost: $0.33 per can

Because $0.33 < $0.34, the better buy is 24 cans for $7.92.

Example 3 Find the more economical purchase: 5 pounds of nails for $3.25, or 4 pounds of nails for $2.58.

Example 4 Find the more economical purchase: 6 cans of fruit for $2.52, or 4 cans of fruit for $1.66.

Strategy To find the more economical purchase, compare the unit costs.

Your strategy

Solution
3.25 ÷ 5 = 0.65
2.58 ÷ 4 = 0.645
$.645 < $.65

The more economical purchase is 4 pounds for $2.58.

Your solution

Solution on p. A24

Objective C To find total cost

An installer of floor tile found the unit cost of identical floor tiles at three stores.

Store 1
$1.22 per tile

Store 2
$1.18 per tile

Store 3
$1.28 per tile

By comparing the unit costs, the installer determined that Store 2 would provide the most economical purchase.

The installer also uses the unit cost to find the total cost of purchasing 300 floor tiles at Store 2. The **total cost** is found by multiplying the unit cost by the number of units purchased.

Unit cost	×	number of units	=	total cost
1.18	×	300	=	354

The total cost is $354.

Example 5 Clear redwood lumber costs $2.43 per foot. How much would 25 feet of clear redwood cost?

Example 6 Pine saplings cost $4.96 each. How much would 7 pine saplings cost?

Strategy To find the total cost, multiply the unit cost ($2.43) by the number of units (25).

Your strategy

Solution

Unit cost	×	number of units	=	total cost
2.43	×	25	=	60.75

The total cost is $60.75.

Your solution

Solution on p. A24

6.1 EXERCISES

▶ Objective A *Application Problems*

Find the unit cost. Round to the nearest tenth of a cent.

1. Syrup, 16 ounces for $0.89
2. Wood, 8 feet for $11.36
3. Lunch meat, 12 ounces for $1.79
4. Light bulbs, 6 bulbs for $2
5. Wood screws, 15 screws for $0.50
6. Bay leaves, $\frac{1}{2}$ ounce for $0.96
7. Antenna wire, 10 feet for $1.95
8. Sponges, 4 for $5.00
9. Catsup, 24 ounces for $0.89
10. Strawberry jam, 18 ounces for $1.23
11. Metal clamps, 2 for $5.99
12. Coffee, 3 pounds for $7.65

▶ Objective B *Application Problems*

Find the more economical purchase.

13. Cheese, 12 ounces for $1.79, or 16 ounces for $2.40
14. Powder, 4 ounces for $0.53, or 9 ounces for $0.95
15. Shampoo, 8 ounces for $1.89, or 12 ounces for $2.64
16. Mayonnaise, 32 ounces for $2.37, or 16 ounces for $1.30
17. Fruit drink, 46 ounces for $0.69, or 50 ounces for $0.78
18. Detergent, 42 ounces for $2.57, or 20 ounces for $1.29

Find the more economical purchase.

19. Oil, 2 quarts for $4.79, or 5 quarts for $12.15

20. Potato chips, 8 ounces for $0.99, or 10 ounces for $1.23

21. Vitamins, 200 tablets for $9.98, or 150 tablets for $7.65

22. Wax, 16 ounces for $2.68, or 27 ounces for $4.39

23. Cereal, 18 ounces for $1.51, or 16 ounces for $1.32

24. Tires, 2 for $128.98, or 4 for $258.98

▶ **Objective C   Application Problems**

Solve.

25. Steak costs $2.79 per pound. Find the total cost of 3 pounds.

26. Red brick costs $0.56 per brick. Find the total cost of 50 bricks.

27. Flowering plants cost $0.49 each. Find the total cost of 6 plants.

28. Chicken costs $0.89 per pound. Find the total cost of 3.6 pounds. Round to the nearest cent.

29. Tea costs $0.54 per ounce. Find the total cost of 6.5 ounces. Round to the nearest cent.

30. Cheese costs $2.89 per pound. Find the total cost of 0.65 pounds. Round to the nearest cent.

31. Tomatoes cost $0.22 per pound. Find the total cost of 1.8 pounds. Round to the nearest cent.

32. Ham costs $1.49 per pound. Find the total cost of 3.4 pounds. Round to the nearest cent.

33. Candy costs $2.68 per pound. Find the total cost of $\frac{3}{4}$ pound.

34. Photocopying costs $0.05 per page. Find the total cost for photocopying 250 pages.

35. Paint costs $8.95 per gallon. How much change do you receive from $50 when purchasing 5 gallons of paint?

36. Crushed rock costs $15.75 per ton. How much change do you receive from $50 when purchasing 3 tons of crushed rock?

SECTION 6.2 Percent Increase and Percent Decrease

Objective A To find percent increase

Percent Increase is used to show how much a quantity has increased over its original value. The statements "car prices will show a 3.5% increase over last year's prices" and "employees were given an 11% pay increase" are illustrations of the use of percent increase.

A city's population increased from 50,000 to 51,500 in 1 year.

To find the percent increase,

first find the amount of increase.

| New value | − | original value | = | amount of increase |

51,500 − 50,000 = 1500

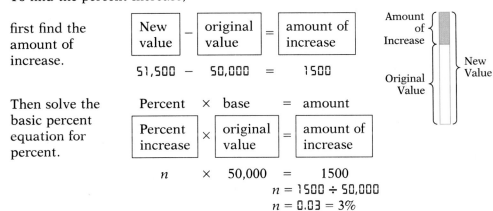

Then solve the basic percent equation for percent.

Percent × base = amount

| Percent increase | × | original value | = | amount of increase |

$n \times 50{,}000 = 1500$
$n = 1500 \div 50{,}000$
$n = 0.03 = 3\%$

The population increased by 3% over the previous year.

Example 1 The average price of gasoline rose from $0.92 to $1.15 in 4 months. What was the percent increase in the price of gasoline?

Strategy To find the percent increase:
- Find the amount of increase.
- Solve the basic percent equation for *percent*.

Solution

| New value | − | original value | = | amount of increase |

1.15 − 0.92 = 0.23

Percent × base = amount
$n \times \$0.92 = \$0.23$
$n = 0.23 \div 0.92$
$n = 0.25 = 25\%$

The percent increase was 25%.

Example 2 An industrial plant increased its number of employees from 2000 to 2250. What was the percent increase in the number of employees?

Your strategy

Your solution

Solution on p. A24

Example 3 A seamstress was making a wage of $4.80 an hour before a 10% increase in pay. What is the new hourly wage?

Strategy To find the new hourly wage:
- Solve the basic percent equation for *amount*.
- Add the amount of increase to the original wage.

Solution Percent × base = amount
$$0.10 \times 4.80 = n$$
$$0.48 = n$$
$$4.80 + 0.48 = 5.28$$

The new hourly wage is $5.28.

Example 4 A baker was making a wage of $6.50 an hour before a 14% increase in pay. What is the new hourly wage?

Your strategy

Your solution

Solution on p. A24

Objective B To apply percent increase to business—markup

Many expenses are involved in operating a business. Employee salaries, rent of office space, and utilities are examples of operating expenses. To pay these expenses and earn a profit, a business must sell a product at a higher price than it paid for the product.

Cost is the price that a business pays for a product, and **selling price** is the price for which a business sells a product to a customer. The difference between selling price and cost is called **markup**.

| Cost | + | markup | = | selling price |

or

| Selling price | − | cost | = | markup |

Markup is frequently expressed as a percent of a product's cost. This percent is called the **markup rate**.

| Markup rate | × | cost | = | markup |

A bicycle store owner purchases a bicycle for $105 and sells it for $147. To find the markup rate,

first find the markup.

| Selling price | − | cost | = | markup |
| 147 | | 105 | | 42 |

Then solve the basic percent equation for *percent*.

Percent × base = amount

| Markup rate | × | cost | = | markup |

$n \times \$105 = \$42$

$n = 42 \div 105 = 0.4$

The markup rate is 40%

Example 5 The manager of a sporting goods store determines that a markup rate of 36% is necessary to make a profit. What is the markup on a pair of skis that costs the store $125?

Strategy To find the markup, solve the basic percent equation for *amount*.

Solution Percent × base = amount

| Markup rate | × | cost | = | markup |

$0.36 \times 125 = n$

$45 = n$

The markup is $45.

Example 6 A bookstore manager determines that a markup rate of 20% is necessary to make a profit. What is the markup on a book that costs the bookstore $8?

Your strategy

Your solution

Example 7 A plant nursery buys a citrus tree for $4.50 and uses a markup-rate of 46%. What is the selling price?

Strategy To find the selling price:
- Find the markup by solving the basic percent equation for *amount*.
- Add the markup to the cost.

Solution Percent × base = amount

| Markup rate | × | cost | = | markup |

$0.46 \times 4.50 = n$

$2.07 = n$

| Cost | + | markup | = | selling price |

$4.50 + 2.07 = 6.57$

The selling price is $6.57.

Example 8 A clothing store buys a suit for $72 and uses a markup rate of 55%. What is the selling price?

Your strategy

Your solution

Solutions on p. A24

Objective C — To find percent decrease

Percent decrease is frequently used to show how much a quantity has decreased from its original value. The statements "the unemployment rate decreased by 0.4% over last month" and "there has been a 12% decrease in the number of industrial accidents" are illustrations of the use of percent decrease.

A family's electric bill decreased from $60 per month to $52.80 per month.

To find the percent decrease, first find the amount of decrease.

| Original value | − | new value | = | amount of decrease |

$$60 - 52.80 = 7.20$$

Then solve the basic percent equation for *percent*.

Percent × base = amount

| Percent decrease | × | original value | = | amount of decrease |

$$n \times \$60 = \$7.20$$
$$n = 7.20 \div 60$$
$$n = 0.12 = 12\%$$

The electric bill decreased by 12% per month.

Example 9 Because of unusually high temperatures, the number of people visiting a desert resort dropped from 650 to 525. Find the percent decrease in attendance. Round to the nearest tenth of a percent.

Strategy To find the percent decrease:
- Find the amount of decrease.
- Solve the basic percent equation for *percent*.

Solution

| Original value | − | new value | = | amount of decrease |

$$650 - 525 = 125$$

Percent × base = amount
$$n \times 650 = 125$$
$$n = 125 \div 650$$
$$n \approx 0.192 = 19.2\%$$

The percent decrease was 19.2%

Example 10 Car sales at an automobile dealership dropped from 150 in June to 120 in July. Find the percent decrease in car sales.

Your strategy

Your solution

Solution on pp. A24–A25

Section 6.2 / Percent Increase and Percent Decrease   227

Example 11 The total sales for December for a stationery store were $16,000. For January, total sales showed an 8% decrease from December's sales. What were the total sales for January?

Example 12 Fog decreased the normal 5-mile visibility at an airport by 40%. What was the visibility in the fog?

Strategy To find the total sales for January:
- Find the amount of decrease by solving the basic percent equation for *amount*.
- Subtract the amount of decrease from the December sales.

Your strategy

Solution Percent × base = amount
$0.08 \times 16{,}000 = n$
$1280 = n$
$16{,}000 - 1280 = 14{,}720$

The total sales for January were $14,720.

Your solution

Solution on pp. A24–A25

Objective D   **To apply percent decrease to business—discount**

To promote sales, a store may reduce the regular price of some of its products temporarily. The reduced price is called the **sale price.** The difference between the regular price and the sale price is called the **discount.**

Discount is frequently stated as a percent of a product's regular price. This percent is called the **discount rate.**

| Discount rate | × | regular price | = | discount |

Example 13
An appliance store has a washing machine that regularly sells for $350 on sale for $297.50. What is the discount rate?

Strategy To find the discount rate:
- Find the discount.
- Solve the basic percent equation for *percent*.

Solution

| Regular price | − | sale price | = | discount |

350 − 297.50 = 52.50

Percent × base = amount

| Discount rate | × | regular price | = | discount |

n × $350 = $52.50
$n = 52.50 \div 350$
$n = 0.15 = 15\%$

The discount rate is 15%.

Example 14
A lawn mower that regularly sells for $125 is on sale for $25 off the regular price. What is the discount rate?

Your strategy

Your solution

Example 15
A plumbing store is selling sinks that regularly sell for $25 at 15% off the regular price. What is the sale price?

Strategy To find the sale price:
- Find the discount by solving the basic percent equation for *amount*.
- Subtract to find the sale price.

Solution

Percent × base = amount

| Discount rate | × | regular price | = | discount |

0.15 × 25 = n
3.75 = n

| Regular price | − | discount | = | sale price |

25 − 3.75 = 21.25

The sale price is $21.25.

Example 16
A florist is selling potted plants that regularly sell for $10.25 at 20% off the regular price. What is the sale price?

Your strategy

Your solution

Solutions on p. A25

6.2 EXERCISES

▶ Objective A *Application Problems*

1. A utility stock that sold for $30 per share increased in price by $1.65 in 1 day. What percent increase does this amount represent?

2. The value of a mutual fund investment of $2500 increased $500. What percent increase does this amount represent?

3. A professional baseball player had a batting average of 0.275. The next year the player had a batting average of 0.297. What is the percent increase in the player's batting average?

4. A running back for a high school football team had a rushing average of 3.80 yards per carry. The next year the back averaged 4.75 yards per carry. What percent increase does this amount represent?

5. You have increased your 60 words per minute typing speed 20 words per minute. What percent increase does this amount represent?

6. Your reading speed increased 70 words per minute from 280 words per minute. What is the percent increase in reading speed?

7. A teacher's union negotiated a new contract calling for a 9% increase from this year's contract.
   a. Find the amount of increase for a teacher who had a salary of $32,500.
   b. Find the new salary for the teacher who had a salary of $32,500.

8. A new labor contract called for an 8.5% increase in pay for all employees.
   a. What is the amount of the increase for an employee who makes $324 per week?
   b. What is the weekly wage of this employee after the wage increase?

9. A company whose stock is selling for $38 per share plans to raise its $1.60 dividend 15%.
   a. What is the increase in the dividend?
   b. Find the total dividend per share of the stock.

10. An amusement park increased its summer staff from 500 to 625 employees.
    a. How many employees were added to the staff?
    b. What percent increase does this amount represent?

11. A family increased its 1440-square foot home $16\frac{2}{3}\%$ by adding a family room. How much larger, in square feet, is the home now?

12. A restaurant increased its weekly hours of operation from 60 hours to 70 hours. What percent increase does this amount represent?

▶ **Objective B** *Application Problems*

13. An air conditioner costs a retail store $285. Find the markup on an air conditioner that has a markup rate of 25%.

14. The owner of an electronics store uses a markup rate of 42%. What is the markup on a cassette player that costs the store $85?

15. A sporting goods store uses a markup rate of 30%. What is the markup on a tennis racket that costs the store $42?

16. The manager of a toy store determines that a markup rate of 38% is necessary for a profit to be made. What is the markup on a novelty item that costs the manager $5?

17. The markup on a computer system that costs $3250 is $975. What markup rate does this amount represent?

18. The markup on a calculator that costs an office supply store $24 is $12. What markup rate does this amount represent?

19. A camera shop uses a markup rate of 48% on its Model ZA cameras, which cost the shop $162.
    a. What is the markup?
    b. What is the selling price?

20. A golf pro shop uses a markup rate of 45% on a set of golf clubs that costs the shop $210.
    a. What is the markup?
    b. What is the selling price?

21. An outdoor furniture company uses a markup rate of 35% on a piece of lawn furniture that costs the store $62.
    a. What is the markup?
    b. What is the selling price?

22. The owner of a roadside fruit stand uses a 55% markup rate and pays $0.60 for a box of strawberries.
    a. What is the markup?
    b. What is the selling price for a box of strawberries?

23. The manager of a building supply store uses a markup rate of 42% for a table saw that costs the store $160. What is the selling price of the table saw?

24. The manager of a magic store uses a markup rate of 48%. What is the selling price of an item that costs $50?

▶ Objective C **Application Problems**

25. A new bridge over a river reduced the normal 45-minute time between two cities by 18 minutes. What percent decrease does this amount represent?

26. In an election, 25,400 voters went to the polls to elect a mayor. Four years later, 1778 fewer voters participated in the election for mayor. What was the percent decrease in the number of voters who went to the polls?

27. By installing energy-saving equipment, a youth club reduced its normal $800 per month utility bill $320. What percent decrease does this amount represent?

28. During the last 40 years, the consumption of eggs in the United States has dropped from 400 eggs per person per year to 260 eggs per person per year. What percent decrease does this amount represent?

29. It is estimated that the value of a new car is reduced 30% after 1 year of ownership. Using this estimate, how much value does a $11,200 new car lose after 1 year?

30. A department store employs 1200 people during the holiday season. At the end of the holiday season, the department store reduces the number of employees 45%. What is the decrease in the number of employees?

31. Because of a decrease in demand for super-8 video cameras, a dealer reduced the orders for these models from 20 per month to 8 per month.
    a. What is the amount of the decrease?
    b. What percent decrease does this amount represent?

32. A new computer reduced the time for printing the payroll from 52 minutes to 39 minutes.
    a. What is the amount of decrease?
    b. What percent decrease does this amount represent?

33. An account executive's average expense for gasoline was $76. After joining a car pool, the executive was able to reduce this expense 20%.
    a. What was the amount of decrease?
    b. What is the average monthly gasoline bill now?

34. Last year a company paid a dividend of $3.50 per share. This year, because of unexpected losses, the company reduced the dividend 12%.
    a. What was the amount of decrease?
    b. What is the dividend this year?

35. Because of an improved traffic pattern at a sports stadium, the average amount of time a fan waits to park decreased from 3.5 minutes to 2.8 minutes. What percent decrease does this amount represent?

36. The price of gold dropped from approximately $600 per ounce to $375 per ounce in 7 months. What percent decrease does this amount represent?

▶ Objective D **Application Problems**

37. A college bookstore is giving a discount of $8 on calculators that normally sell for $24. What is the discount rate?

38. A clothing store is selling a $72 sport jacket for $24 off the regular price. What is the discount rate?

39. A disk player that regularly sells for $340 is selling for 20% off the regular price. What is the discount?

40. An appliance store is selling its $450 washing machines for 15% off the regular price. What is the discount?

41. To promote business, a store manager offers an electric grill that regularly sells for $140 at $42 off the regular price. What is the discount rate?

42. A computer system that regularly sells for $1400 is on sale for $350 off the regular price. What is the discount rate?

43. A service station has its regularly priced $45 tune-up on sale for 16% off the regular price.
    a. What is the discount?
    b. What is the sale price?

44. Turkey that regularly sells for $0.85 per pound is on sale for 20% off the regular price.
    a. What is the discount?
    b. What is the sale price?

45. A department store has regularly priced $160 sleeping bags on sale for $120.
    a. What is the discount?
    b. What is the discount rate?

46. A nationally advertised brand of paint that regularly costs $16 per gallon is on sale for $12 per gallon.
    a. What is the discount?
    b. What is the discount rate?

47. A clothing store is offering 15% off its entire stock of children's clothing. What is the sale price of a jacket that regularly sells for $42?

48. During a sidewalk sale, all merchandise was reduced 20% off the regular price. What was the sale price of a sofa that normally sold for $650?

SECTION 6.3 Interest

Objective A   To calculate simple interest

When money is deposited in a bank account, the bank pays the depositor for the privilege of using that money. The amount paid to the depositor is called **interest.** When money is borrowed from a bank, the borrower pays for the privilege of using that money. The amount paid to the bank is also called interest.

The original amount deposited or borrowed is called the **principal.** The amount of interest is a percent of the principal. The percent used to determine the amount of interest is the **interest rate.** Interest rates are given for specific periods of time, usually months or years.

Interest computed on the original principal is called **simple interest.** To calculate simple interest, multiply the principal by the interest rate per period by the number of time periods.

Calculate the simple interest on $1500 deposited for 2 years at an annual interest rate of 7.5%. As shown below, the interest earned is $225.

Principal × annual interest rate × time (in years) = interest
$$1500 \times 0.075 \times 2 = 225$$

Example 1
A person borrows $500 from a savings and loan association for 6 months at an annual interest rate of 7%. What is the simple interest due on the loan?

Strategy
To find the simple interest, multiply:

Principal × annual   × time
          interest rate (in years)

Solution $500 \times 0.07 \times 0.5 = 17.5$

The interest due is $17.50.

Example 2
A company borrows $15,000 from a bank for 18 months at an annual interest rate of 12%. What is the simple interest due on the loan?

Your strategy

Your solution

Example 3
A credit card company charges a customer 1.5% per month on the unpaid balance of charges on the credit card. What is the interest due in a month when the customer has an unpaid balance of $54?

Strategy
To find the interest due, multiply the principal by the monthly interest rate by the time (in months).

Solution $54 \times 0.015 \times 1 = 0.81$

The interest charge is $0.81.

Example 4
A bank offers short-term loans at an interest rate of 2% per month. What is the interest due on a short-term loan of $400 for 2 months?

Your strategy

Your solution

Solutions on p. A25

Objective B To calculate compound interest

Usually the interest paid on money deposited or borrowed is compound interest. **Compound interest** is computed not only on the original principal but also on interest already earned. Compound interest is usually compounded annually (once a year), semiannually (twice a year), quarterly (four times a year), or daily.

$100 is invested for 3 years at an annual interest rate of 9%, compounded annually. The interest earned over the 3 years is calculated by first finding the interest earned each year.

1st year	Interest earned:	0.09 × $100.00 = $9.00
	New principal:	$100.00 + $9.00 = $109.00
2nd year	Interest earned:	0.09 × $109.00 = $9.81
	New principal:	$109.00 + $9.81 = $118.81
3rd year	Interest earned:	0.09 × $118.81 ≈ $10.69
	New principal:	$118.81 + $10.69 = $129.50

To find the interest earned, subtract the original principal from the new principal. As shown below, the interest earned is $29.50.

New principal	−	original principal	=	interest earned
129.50	−	100	=	29.5

Note that the compound interest earned is $29.50, and the simple interest earned on the investment would have been only $100 × 0.09 × 3 = $27.

Calculating compound interest can be very tedious, so there are tables that can be used to simplify these calculations. A portion of a compound interest table is given on pages A4 and A5.

Example 5
An investment of $650 pays 8% annual interest compounded semiannually. What is the interest earned in 5 years?

Strategy To find the interest earned:
- Find the new principal by multiplying the original principal by the factor found in the compound interest table.
- Subtract the original principal from the new principal.

Solution
650 × 1.48024 ≈ 962.16
          ↓   ↓
      from the   new
       table   principal
962.16 − 650 = 312.16

The interest earned is $312.16.

Example 6
An investment of $1000 pays 10% annual interest compounded quarterly. What is the interest earned in 20 years?

Your strategy

Your solution

Solution on p. A25

6.3 EXERCISES

▶ **Objective A** *Application Problems*

1. To finance the purchase of 15 new cars, a car rental agency borrows $100,000 for 9 months at an annual interest rate of 12%. What is the simple interest due on the loan?

2. A home builder borrows $50,000 for 8 months at an annual interest rate of 11.5%. What is the simple interest due on the loan? (Round to the nearest cent.)

3. A farmer obtained a $25,000 loan at a 13.5% annual interest rate for 4 years. Find the simple interest due on the loan.

4. A nursery operator borrowed $150,000 at a 9.5% annual interest rate for 4 years. What is the simple interest due on the loan?

5. A credit union charges its customers an interest rate of 2% per month for transferring money into an account that is overdrawn. Find the interest owed to the bank for 1 month when $800 was transferred into an overdrawn account.

6. A credit card company charges a customer 1.6% per month on the customer's unpaid balance. Find the interest owed to the credit card company when the customer's unpaid balance for the month is $1250.

7. A metalworking shop purchased a robot-controlled lathe for $225,000 and financed the full amount at 8% simple annual interest for 4 years.
   a. Find the interest due on the loan.
   b. Find the monthly payment. (Monthly payment = $\frac{\text{loan amount} + \text{interest}}{\text{number of months}}$)

8. An entertainment center is purchased, and an $1800 loan is obtained for 2 years at a simple interest rate of 12.4%.
   a. Find the interest due on the loan.
   b. Find the monthly payment. (Monthly payment = $\frac{\text{loan amount} + \text{interest}}{\text{number of months}}$)

9. To attract new customers, a car dealership is offering car loans at a simple interest rate of 4.5%.
   a. Find the interest charged to a customer who finances a car loan of $12,000 for 2 years.
   b. Find the monthly payment.

10. A real estate syndicator purchases a small plane for $57,000 and finances the full amount for 5 years at a simple annual interest rate of 10.5%.
    a. Find the interest due on the loan.
    b. Find the monthly payment.

11. A homeowner decided to build onto an existing structure instead of buying a new home. The homeowner borrowed $42,000 for $3\frac{1}{2}$ years at a simple interest rate of 9.5%. Find the monthly payment.

12. An office has 15 new computers installed for a total cost of $24,000. The entire amount was financed for $2\frac{1}{2}$ years at a simple interest rate of 11.2%. Find the monthly payment.

▶ Objective B **Application Problems**

Solve. Use the table on pages A4 and A5. Round to the nearest cent.

13. A credit union pays 9% annual interest, compounded daily, on time savings deposits. Find the value of $750 deposited in this account after 1 year.

14. A high school student has an investment of $1000 that pays 9% annual interest, compounded quarterly. What is the value of the investment after 5 years?

15. An investment club invests $50,000 in a certificate of deposit that pays 10% annual interest, compounded quarterly. Find the value of this investment after 10 years.

16. A teacher invests $2500 in a tax-sheltered annuity that pays 8% annual interest, compounded daily. Find the value of this investment after 20 years.

17. A computer specialist invests $3000 in a corporate retirement account that pays 9% annual interest compounded semiannually. Find the value of this investment after 15 years.

18. To replace equipment, a farmer invests $20,000 in an account that pays 7% annual interest, compounded semiannually. What is the value of the investment after 5 years?

19. A corporation invests $75,000 in a trust account that pays 8% interest, compounded quarterly.
    a. What will the value of the investment be in 5 years?
    b. How much interest will be earned in the 5 years?

20. To save for retirement, a couple deposited $3000 in an account that pays 11% annual interest, compounded daily.
    a. What will the value of the investment be in 10 years?
    b. How much interest will be earned in the 10 years?

21. A fast-food restaurant has determined that it will need $50,000 in 5 years for expansion. In an attempt to meet that goal, the restaurant invests $35,000 today in an account that pays 8% annual interest, compounded quarterly.
    a. Find the value of the $35,000 investment in 5 yrs.
    b. Will the restaurant have enough money in the account for the planned expansion?

22. An investment of $5000 in an account that pays 10% annual interest, compounded quarterly, has a value of $8193.10 after 5 years.
    a. What would the value of the investment have been if the 10% interest had been compounded daily over the 5 year period?
    b. What is the difference in the values of the investment after the 5 years?

23. To save for a child's college education, a family deposits $2500 into an account that pays 10% annual interest, compounded daily. Find the amount of interest earned in this account over a 20-year period.

24. An account executive suggests buying a $5000 municipal bond that pays 8% annual interest, compounded quarterly. Find the amount of interest that would be earned in 5 years.

SECTION 6.4   Real Estate Expenses

Objective A   To calculate the initial expenses of buying a home

One of the largest investments a person will make is the purchase of a home. The major initial expense in the purchase is the down payment. The amount of the down payment is normally a percent of the purchase price. This percent varies among banks, but it usually ranges from 5% to 25%.

The **mortgage** is the amount that is borrowed to buy real estate. The mortgage amount is the difference between the purchase price and the down payment.

A home is purchased for $85,000, and a down payment of $12,750 is made. Find the mortgage.

| Purchase price | − | down payment | = | mortgage |

$$85{,}000 \;-\; 12{,}750 \;=\; 72{,}250 \qquad \text{The mortgage is \$72,250.}$$

Another large initial expense in buying a home is the loan origination fee, which is a fee the bank charges for processing the mortgage papers. The loan origination fee is usually a percent of the mortgage and is expressed in **points,** which is the term banks use to mean percent. For example, "5 points" means "5 percent."

| Points | × | mortgage | = | loan origination fee |

Example 1

A house is purchased for $87,000, and a down payment, which is 20% of the purchase price, is made. Find the mortgage.

Strategy

To find the mortgage:
- Find the down payment by solving the basic percent equation for *amount*.
- Subtract the down payment from the purchase price.

Solution

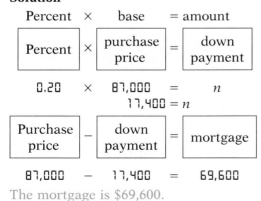

The mortgage is $69,600.

Example 2

An office building is purchased for $216,000, and a down payment, which is 25% of the purchase price, is made. Find the mortgage.

Your strategy

Your solution

Solution on p. A26

Example 3 A home is purchased with a mortgage of $65,000. The buyer pays a loan origination fee of $3\frac{1}{2}$ points. How much is the loan origination fee?

Example 4 The mortgage on a real estate investment is $80,000. The buyer paid a loan origination fee of $4\frac{1}{2}$ points. How much is the loan origination fee?

Strategy To find the loan origination fee, solve the basic percent equation for *amount*.

Your strategy

Solution
Percent × base = amount

| Points | × | mortgage | = | fee |

0.035 × 65,000 = n
2275 = n

The loan origination fee is $2275.

Your solution

Solution on p. A26

Objective B To calculate ongoing expenses of owning a home

Besides the initial expenses of buying a home, there are continuing monthly expenses involved in owning a home. The monthly mortgage payment, utilities, insurance, and taxes are some of these ongoing expenses. Of these expenses, the largest one is normally the monthly mortgage payment.

For a fixed rate mortgage, the monthly mortgage payment remains the same throughout the life of the loan. The calculation of the monthly mortgage payment is based on the amount of the loan, the interest rate on the loan, and the number of years required to pay back the loan. Calculating the monthly mortgage payment is fairly difficult, so tables such as the one on page A6 are used to simplify these calculations.

Find the monthly mortgage payment on a 30-year $60,000 mortgage at an interest rate of 9%. Use the monthly payment table on page A6.

60,000 × 0.0080462 = 482.77
                ↓
            from the
             table

The monthly mortgage payment is $482.77.

The monthly mortgage payment includes the payment of both principal and interest on the mortgage. The interest charged during any one month is charged on the unpaid balance of the loan. Therefore, during the early years of the mortgage, when the unpaid balance is high, most of the monthly mortgage payment is interest charged on the loan. During the last few years of a mortgage, when the unpaid balance is low, most of the monthly mortgage payment goes toward paying off the loan.

Find the interest paid on a mortgage during a month when the monthly mortgage payment is $186.26 and $58.08 of that amount goes toward paying off the principal.

| Monthly mortgage payment | − | principal | = | interest |

$$186.26 - 58.08 = 128.18$$

The interest paid on the mortgage is $128.18.

Property tax is another ongoing expense of owning a house. Property tax is normally an annual expense that is paid on a monthly basis. The monthly property tax, which is determined by dividing the annual property tax by 12, is usually added to the monthly mortgage payment.

A home owner must pay $534 in property tax annually. Find the property tax that must be added each month to the home owner's monthly mortgage payment.

$$534 \div 12 = 44.5$$

$44.50 must be added to the monthly mortgage payment each month for property tax.

Example 5

A farmer purchases some land for $120,000 and makes a down payment of $25,000. The savings and loan association charges an annual interest rate of 11% on the farmer's 25-year mortgage. Find the monthly mortgage payment.

Strategy

To find the monthly mortgage payment:
- Subtract the down payment from the purchase price to find the mortgage.
- Multiply the mortgage by the factor found in the monthly payment table on page A6.

Solution

| Purchase price | − | down payment | = | mortgage |

$$120,000 - 25,000 = 95,000$$

$$95,000 \times \underset{\text{from the table}}{0.0098011} = 931.10$$

The monthly mortgage payment is $931.10.

Example 6

A new condominium project is selling townhouses for $75,000. A down payment of $15,000 is required, and a 20-year mortgage at an annual interest rate of 12% is available. Find the monthly mortgage payment.

Your strategy

Your solution

Solution on p. A26

Example 7 A home has a mortgage of $55,000 for 25 years at an annual interest rate of 10%. During a month when $375.88 of the monthly mortgage payment is principal, how much of the payment is interest?

Strategy To find the interest:
- Multiply the mortgage by the factor found in the monthly payment table on page A6 to find the monthly mortgage payment.
- Subtract the principal from the monthly mortgage payment.

Solution 55,000 × 0.0090870 ≈ 499.79
↓ from the table   ↓ monthly mortgage payment

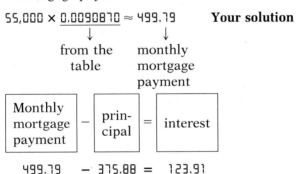

499.79 − 375.88 = 123.91

$123.91 is interest on the mortgage.

Example 8 An office building has a mortgage of $125,000 for 25 years at an annual interest rate of 9%. During a month when $492.65 of the monthly mortgage payment is principal, how much of the payment is interest?

Your strategy

Your solution

Example 9 The monthly mortgage payment for a home is $598.75. The annual property tax is $900. Find the total monthly payment for the mortgage and property tax.

Strategy To find the monthly payment:
- Divide the annual property tax by 12 to find the monthly property tax.
- Add the monthly property tax to the monthly mortgage payment.

Solution 900 ÷ 12 = 75   monthly property tax

598.75 + 75 = 673.75

The total monthly payment is $673.75.

Example 10 The monthly mortgage payment for a home is $415.20. The annual property tax is $744. Find the total monthly payment for the mortgage and property tax.

Your strategy

Your solution

Solutions on p. A26

6.4 EXERCISES

▶ Objective A *Application Problems*

1. A condominium at a ski resort is purchased for $97,000, and a down payment of $14,550 is made. Find the mortgage.

2. An insurance business is purchased for $173,000, and a down payment of $34,600 is made. Find the mortgage.

3. A building lot is purchased for $25,000. The lender requires a down payment of 30%. Find the down payment.

4. A new home buyer purchases a house for $88,500. The lender requires a down payment of 20%. Find the down payment.

5. An investor makes a down payment of 25% of the $250,000 purchase price of an apartment building. How much is the down payment?

6. A clothing store is purchased for $125,000, and a down payment, which is 25% of the purchase price, is made. How much is the down payment?

7. A loan of $150,000 is obtained to purchase a home. The loan origination fee is $2\frac{1}{2}$ points. Find the loan origination fee.

8. A savings and loan requires a borrower to pay $3\frac{1}{2}$ points for a loan. Find the loan origination fee for a loan of $90,000.

9. A real estate developer is selling developed homes for $95,000. A down payment of 5% is required.
   a. Find the down payment.
   b. Find the mortgage.

10. A cattle rancher purchases some land for $240,000. The bank requires a down payment of 15%.
    a. Find the down payment.
    b. Find the mortgage.

11. An electrician purchases a home for $96,000. Find the mortgage if the down payment is 10% of the purchase price.

12. A mortgage lender requires a down payment of 5% of the $56,000 purchase price of a condominium. How much is the mortgage?

Objective B Application Problems

Solve. Use the monthly payment table on page A6. Round to the nearest cent.

13. An investment club obtained a loan of $150,000 to buy a car wash business. The monthly mortgage payment was based on 25 years at 10%. Find the monthly mortgage payment.

14. A beautician obtains a 20-year mortgage of $90,000 to expand the business. The credit union charges an annual interest rate of 11%. Find the monthly mortgage payment.

15. A couple interested in buying a home determines that they can afford a monthly mortgage payment of $500. Can they afford to buy a home with a 30-year $65,000 mortgage at 9% interest?

16. A lawyer is considering purchasing a new office building with a 20-year $400,000 mortgage at 12% interest. The lawyer can afford a monthly mortgage payment of $6000. Can the lawyer afford the monthly mortgage payments on the new office building?

17. The county tax assessor has determined that the annual property tax on a $58,000 house is $645. Find the monthly property tax.

18. The annual property tax on an $86,000 home is $858. Find the monthly property tax.

19. An import company has a warehouse with a 25-year mortgage of $200,000 at an annual interest rate of 11%.
    a. Find the monthly mortgage payment.
    b. During a month when $941.72 of the monthly mortgage payment is principal, how much of the payment is interest?

20. A vacation home has a mortgage of $72,000 for 30 years at an annual interest rate of 10%.
    a. Find the monthly mortgage payment.
    b. During a month when $392.47 of the monthly mortgage payment is principal, how much of the payment is interest?

21. A duplex has a monthly mortgage payment of $452.73. The owner must pay an annual property tax of $948. Find the total monthly payment for the mortgage and property tax.

22. The monthly mortgage payment on a home is $356.48. The home owner must pay an annual property tax of $792. Find the total monthly payment for the mortgage and property tax.

23. A health inspector purchased a home for $86,000 and made a down payment of $15,000. The balance was financed for 30 years at an annual interest rate of 11%. Find the monthly mortgage payment.

24. A customer of a savings and loan purchased a $75,000 home and made a down payment of $20,000. The savings and loan charges its customers an annual interest rate of 9% for 30 years for a home mortgage. Find the monthly mortgage payment.

SECTION 6.5 Car Expenses

Objective A To calculate the initial expenses of buying a car

The initial expenses in the purchase of a car usually include the down payment, the license fees, and the sales tax. The down payment may be very small or as much as 25 or 30 percent of the purchase price of the car, depending on the lending institution. License fees and sales tax are regulated by each state, so these expenses vary from state to state.

Example 1
A car is purchased for $8500, and the lender requires a down payment of 15% of the purchase price. Find the amount financed.

Strategy
To find the amount financed:

- Find the down payment by solving the basic percent equation for *amount*.
- Subtract the down payment from the purchase price.

Solution

Percent × base = amount

Percent × purchase price = down payment

0.15 × 8500 = n

1275 = n

8500 − 1275 = 7225

The amount financed is $7225.

Example 2
A down payment of 20% of the $9200 purchase price of a new car is made. Find the amount financed.

Your strategy

Your solution

Example 3
A sales clerk purchases a car for $6500 and pays a sales tax that is 5% of the purchase price. How much is the sales tax?

Strategy
To find the sales tax, solve the basic percent equation for *amount*.

Solution

Percent × base = amount

Percent × purchase price = sales tax

0.05 × 6500 = n

325 = n

The sales tax is $325.

Example 4
A car is purchased for $7350. The car license fee is 1.5% of the purchase price. How much is the license fee?

Your strategy

Your solution

Solutions on pp. A26–A27

Objective B To calculate ongoing expenses of owning a car

Besides the initial expenses of buying a car, there are continuing expenses involved in owning a car. These ongoing expenses include car insurance, gas and oil, general maintenance, and monthly car payment. The monthly car payment is calculated in the same manner as monthly mortgage payments on a home loan. A monthly payment table, such as the one on page A6, is used to simplify calculation of monthly car payments.

Example 5
At a cost of $0.27 per mile, how much does it cost to operate a car during a year the car is driven 15,000 miles?

Strategy
To find the cost, multiply the cost per mile by the number of miles driven.

Solution
15,000 × 0.27 = 4050   The cost is $4050.

Example 6
At a cost of $0.22 per mile, how much does it cost to operate a car during a year the car is driven 23,000 miles?

Your strategy

Your solution

Example 7
During 1 month the total gasoline bill was $84 and the car was driven 1200 miles. What was the cost per mile for gasoline?

Strategy
To find the cost per mile for gasoline, divide the cost for gasoline by the number of miles driven.

Solution   84 ÷ 1200 = 0.07
The cost per mile was $0.07.

Example 8
In a year when the total car insurance bill was $360 and the car was driven 15,000 miles, what was the cost per mile for car insurance?

Your strategy

Your solution

Example 9
A car is purchased for $8500 with a down payment of $1700. The balance is financed for 3 years at an annual interest rate of 12%. Find the monthly car payment.

Strategy
To find the monthly payment:
- Subtract the down payment from the purchase price to find the amount financed.
- Multiply the amount financed by the factor found in the monthly payment table on page A6.

Solution
8500 − 1700 = 6800
            ↓
     amount financed

6800 × 0.0332143 ≈ 225.86

The monthly payment is $225.86.

Example 10
A truck is purchased for $15,900 with a down payment of $3975. The balance is financed for 4 years at an annual interest rate of 11%. Find the monthly payment.

Your strategy

Your solution

Solutions on p. A27

6.5 EXERCISES

▶ Objective A *Application Problems*

1. A receptionist has saved $780 to make a down payment on a car. The bank requires a down payment of 12% of the purchase price of $7100. Does the receptionist have enough money to make the down payment on the car?

2. A jeep is purchased for $9288. A down payment of 15% is required. How much is the down payment?

3. A drapery installer buys a minivan to carry drapery samples. The purchase price of the van is $16,500, and a 4.5% sales tax is paid. How much is the sales tax?

4. A delivery truck for a lumber yard is purchased for $18,500. A sales tax of 4% of the purchase price is paid. Find the sales tax.

5. A license fee of 2% of the purchase price of a truck is to be paid on a pickup truck costing $12,500. How much is the license fee for the truck?

6. Your state charges a car license fee of 1.5% of the purchase price of a car. How much is the license fee for a car that costs $6998?

7. An electrician buys a $12,000 flat-bed truck. A state license fee of $175 and a sales tax of 3.5% of the purchase price are required.
   a. Find the sales tax.
   b. Find the total cost of the sales tax and the license fee.

8. A physical therapist buys a car for $9375 with a down payment of $1875 and a sales tax of 5% of the purchase price.
   a. Find the sales tax.
   b. Find the total cost of the sales tax and the down payment.

9. A student buys a motorcycle for $2200 and makes a down payment of 25% of the purchase price.
   a. Find the down payment.
   b. Find the amount financed.

10. A carpenter buys a utility van for $14,900 and makes a down payment that is 15% of the purchase price.
    a. Find the down payment.
    b. Find the amount financed.

11. An author buys a sports car for $35,000 and makes a down payment of 20% of the purchase price. Find the amount financed.

12. An executive secretary purchases a new car for $13,500. The secretary makes a down payment of 25% of the cost. Find the amount financed.

▶ Objective B   *Application Problems*

Solve. Use the monthly payment table on page A6. Round to the nearest cent.

13. A rancher purchases a truck, and $14,000 is financed through a credit union at 11% interest for 4 years. Find the monthly truck payment.

14. A car loan of $8000 is financed through a bank at an annual interest rate of 10% for 3 years. Find the monthly car payment.

15. An estimate of the cost and care of owning a compact car is $0.32 per mile. Using this estimate, how much does it cost to operate a car during a year the car is driven 16,000 miles?

16. An estimate of the cost of care and maintenance of automobile tires is $0.015 per mile. Using this estimate, how much would it cost for care and maintenance during a year the car is driven 14,000 miles?

17. A family spent $1600 on gas, oil, and car insurance during a period when the car was driven 14,000 miles. Find the cost per mile for gas, oil, and car insurance.

18. Last year, you spent $1050 for gasoline for your car. The car was driven 15,000 miles. What was the cost per mile for gasoline?

19. A car owner pays a monthly car payment of $143.50. During a month when $68.75 of the monthly payment is principal, how much of the payment is interest?

20. The cost for a pizza delivery truck for the year included $1870 in truck payments, $1200 on gasoline, and $675 on insurance. Find the total cost for truck payments, gasoline, and insurance for the year.

21. A city purchases a fire truck for $36,000, and a down payment of $5400 is made. The balance is financed for 5 years at an annual rate of 10%.
    a. Find the amount financed.
    b. Find the monthly truck payment.

22. A used car is purchased for $4995, and a down payment of $995 is made. The balance is financed for 3 years at an interest rate of 11%.
    a. Find the amount financed.
    b. Find the monthly car payment.

23. An artist purchases a new car costing $19,500, with a down payment of $3900. The balance is financed for 3 years at an annual interest rate of 10%. Find the monthly car payment.

24. A half-ton truck with a camper is purchased for $12,500, and a down payment of $1500 is made. The balance is financed for 4 years at an annual interest rate of 11%. Find the monthly payment.

SECTION 6.6 Wages

Objective A To calculate commissions, total hourly wages, and salaries

Commissions, hourly wage, and salary are three forms of receiving payment for doing work.

Commissions are usually paid to salespersons and are calculated as a percent of total sales.

A real estate broker receives a commission of 4.5% of the selling price of a house. Find the commission the broker earned for selling a home for $75,000.

To find the commission the broker earned, solve the basic percent equation for *amount*.

Percent × base = amount

| Commission rate | × | total sales | = | commission |

0.045 × 75,000 = 3375

The commission is $3375.

An employee who receives an **hourly wage** is paid a certain amount for each hour worked.

A plumber receives an hourly wage of $13.25. Find the plumber's total wages for working 37 hours.

To find the plumber's total wages, multiply the hourly wage by the number of hours worked.

| Hourly wage | × | number of hours worked | = | total wages |

13.25 × 37 = 490.25

The plumber's total wages for working 37 hours are $490.25.

An employee who is paid a **salary** receives payment based on a weekly, biweekly (every other week), monthly, or annual time schedule. Unlike the employee who receives an hourly wage, the salaried worker does not receive additional pay for working more than the regularly scheduled work day.

A computer operator receives a weekly salary of $295. Find the computer operator's salary for 1 month (4 weeks).

To find the computer operator's salary for 1 month, multiply the salary per pay period by the number of pay periods.

| Salary per pay period | × | number of pay periods | = | total salary |

295 × 4 = 1180

The computer operator's total salary for 1 month is $1180.

Example 1
A pharmacist's hourly wage is $18. On Saturday, the pharmacist earns time and a half ($1\frac{1}{2}$ times the regular hourly wage). How much does the pharmacist earn for working 6 hours on Saturday?

Strategy
To find the pharmacist's earnings:
- Find the hourly wage for working on Saturday by multiplying the hourly wage by $1\frac{1}{2}$.
- Multiply the product by the number of hours worked.

Solution
18 × 1.5 = 27   27 × 6 = 162

The pharmacist earns $162.

Example 2
A construction worker's hourly wage is $8.50. The worker earns double time (2 times the regular hourly wage) for working overtime. How much does the worker earn for working 8 hours of overtime?

Your strategy

Your solution

Example 3
An efficiency expert received a contract for $3000. The consultant spent 75 hours on the project. Find the consultant's hourly wage.

Strategy
To find the hourly wage, divide the total earnings by the number of hours worked.

Solution
3000 ÷ 75 = 40   The hourly wage was $40.

Example 4
A contractor for a bridge project receives an annual salary of $28,224. What is the contractor's salary per month?

Your strategy

Your solution

Example 5
A computer salesperson receives $18,500 per year plus a $5\frac{1}{2}$% commission on sales over $100,000. During the year, the salesperson sold $150,000 worth of computers. Find the salesperson's total earnings for the year.

Strategy
To find the total earnings:
- Find the commission earned by multiplying the commission rate by sales over $100,000.
- Add the commission to the annual pay.

Solution
150,000 − 100,000 = 50,000
50,000 × 0.055 = 2750 commission
18,500 + 2750 = 21,250

The salesperson earned $21,250.

Example 6
An insurance agent receives $12,000 per year plus a $9\frac{1}{2}$% commission on sales over $50,000. During 1 year, the agent's sales totaled $175,000. Find the agent's total earnings for the year.

Your strategy

Your solution

Solutions on p. A27

6.6 EXERCISES

▶ Objective A *Application Problems*

1. A salesclerk in a clothing store earns $5.50 per hour. How much does the clerk earn in a 40-hour week?

2. A homeowner pays a gardener an hourly wage of $5.25. How much does the homeowner pay the gardener for working 25 hours?

3. A real estate agent receives a 3% commission for selling a house. Find the commission the real estate agent earned for selling a house for $98,500.

4. An insurance agent receives a commission of 40% of the first year's premium. Find the agent's commission for selling a life insurance policy with a first year premium of $1050.

5. A stock broker receives a commission of 1.5% of the price of stock that is bought or sold. Find the commission on 100 shares of stock that were bought for $5600.

6. The owner of a gallery receives a commission of 20% of paintings that are sold on consignment. Find the commission on a painting sold for $22,500.

7. A teacher receives an annual salary of $38,928. How much does the teacher receive per month?

8. An apprentice plumber receives an annual salary of $19,680. How much does the plumber receive per month?

9. A car salesperson receives a 4% commission for selling a new car. Find the commission the salesperson earned for selling a new car for $10,400.

10. An electrician's hourly wage is $15.80. For working overtime, the electrician earns double time. What is the electrician's hourly wage for working overtime?

11. A sales representative for a shoe company receives a commission of 12% of the weekly sales. Find the commission earned during the week when sales were $4500.

12. A golf pro receives a commission of 25% for selling a golf set. Find the commission the pro earned for selling a golf set that cost $450.

13. A carpet installer receives $1.75 per square yard to install carpet. How much does the installer receive for installing 160 square yards of carpet?

14. A typist charges $1.75 per page for typing technical material. How much does the typist earn for typing a 225-page book?

15. A nuclear chemist receives $15,000 for consulting fees while working on a nuclear power plant. The chemist worked 120 hours on the project. Find the consultant's hourly wage.

16. A computer consultant received a contract for $3400. The consultant spent 40 hours working on the project. Find the consultant's hourly wage.

17. A physical therapist's hourly wage is $10.78. For working overtime, the therapist receives double time.
    a. What is the therapist's hourly wage for working overtime?
    b. How much does the therapist earn for working 16 hours of overtime?

18. A lathe operator receives an hourly wage of $8.50. When working on Saturday, the operator receives time and a half.
    a. What is the lathe operator's hourly wage on Saturday?
    b. How much does the lathe operator earn for working 8 hours on Saturday?

19. A stock clerk at a supermarket earns an hourly wage of $6.20. For working the late night shift, the clerk's wage is increased by 15%.
    a. What is the increase in pay for working the late night shift?
    b. What is the clerk's hourly wage for working the late night shift?

20. A nurse's hourly wage is $8.50. For working the night shift, the nurse's wage is increased 10%.
    a. What is the increase in pay for working the night shift?
    b. What is the nurse's hourly wage for working the night shift?

21. A sales executive sells heavy equipment to a highway construction company. The executive receives 14% of all sales over $100,000. During 1 year, the sales executive sold $350,000 worth of heavy equipment.
    a. Find the amount of sales over $100,000.
    b. Find the commission the sales executive earned.

22. A sales engineer receives a salary of $15,000 per year plus a commission of 20% of the total annual sales. During 1 year, the total sales were $125,000.
    a. Find the commission the sales engineer earned.
    b. Find the sales engineer's total earnings for the year.

23. A service station attendant's hourly wage is $6.40. For working the night shift, the attendant's wage is increased 15%. What is the attendant's hourly wage for working the night shift?

24. A door-to-door salesperson receives a salary of $150 per week plus a commission of 15% on all sales over $1500. Find the salesperson's earnings during a week that sales totaled $3000.

SECTION 6.7   Bank Statements

Objective A   To calculate checkbook balances

A checking account can be opened at most banks or savings and loan associations by depositing an amount of money in the bank. A checkbook contains checks and deposit slips and a checkbook register in which to record checks written and amounts deposited in the checking account. Each time a check is written, the amount of the check is subtracted from the amount in the account. When a deposit is made, the amount deposited is added to the amount in the account.

A portion of a checkbook register is shown below. The account holder had a balance of $587.93 before writing two checks, one for $286.87 and the other for $102.38, and making one deposit for $345.00.

		RECORD ALL CHARGES OR CREDITS THAT AFFECT YOUR ACCOUNT					BALANCE	
NUMBER	DATE	DESCRIPTION OF TRANSACTION	PAYMENT/DEBIT (-)	√T	FEE (IF ANY) (-)	DEPOSIT/CREDIT (+)	$ 587	93
108	8/4	Plumber	$286 87		$	$	301	06
109	8/10	Car Payment	102 38				198	68
	8/14	Deposit				345 00	543	68

To find the current checking account balance, subtract the amount of each check from the previous balance. Then add the amount of the deposit.

The current checking account balance is $543.68.

Example 1 A mail carrier had a checking account balance of $485.93 before writing two checks, one for $18.98 and another for $35.72, and making a deposit of $250. Find the current checking account balance.

Example 2 A cement mason had a checking account balance of $302.46 before writing a check for $20.59 and making two deposits, one in the amount of $176.86 and another in the amount of $94.73. Find the current checking account balance.

Strategy To find the current balance:
- Subtract the amounts of each check from the old balance.
- Add the amount of the deposit.

Your strategy

Solution

```
   485.93
 -  18.98   first check
   466.95
 -  35.72   second check
   431.23
 + 250.00   deposit
   681.23
```

The current checking account balance is $681.23.

Your solution

Solution on p. A28

Objective B — To balance a checkbook

Each month a bank statement is sent to the account holder. The bank statement shows the checks that the bank has paid, the deposits received, and the current bank balance.

A bank statement and checkbook register are shown on the next page.

Balancing the checkbook, or determining if the checking account balance is accurate, requires a number of steps.

1. In the checkbook register, put a check (√) by each check paid by the bank and each deposit recorded by the bank.

Section 6.7 / Bank Statements   **253**

NUMBER	DATE	DESCRIPTION OF TRANSACTION	PAYMENT/DEBIT (−)	√ T	FEE (IF ANY) (−)	DEPOSIT/CREDIT (+)	BALANCE $ 840 27
263	5/20	Dentist	$ 25 00	√	$	$	815 27
264	5/21	Meat market	33 61	√			781 66
265	5/22	Gas company	67 14				714 52
	5/29	Deposit		√		192 00	906 52
266	5/29	Pharmacy	18 95	√			887 57
267	5/30	Telephone	43 85				843 72
268	6/2	Groceries	43 19	√			800 53
	6/3	Deposit		√		215 00	1015 53
269	6/7	Insurance	103 00	√			912 53
	6/10	Deposit				225 00	1137 53
270	6/15	Clothing store	16 63	√			1120 90
271	6/18	Newspaper	7 00				1113 90

CHECKING ACCOUNT Monthly Statement   Account Number: 924-297-8

Date	Transaction	Amount	Balance
5/20	OPENING BALANCE		840.27
5/21	CHECK	25.00	815.27
5/23	CHECK	33.61	781.66
5/29	DEPOSIT	192.00	973.66
6/1	CHECK	18.95	954.71
6/1	INTEREST	4.47	959.18
6/3	CHECK	43.19	915.99
6/3	DEPOSIT	215.00	1130.99
6/9	CHECK	103.00	1027.99
6/16	CHECK	16.63	1011.36
6/20	SERVICE CHARGE	3.00	1008.36
6/20	CLOSING BALANCE		1008.36

2. Add to the current checkbook balance all checks that have been written but not yet paid by the bank and any interest paid on the account.

3. Subtract any service charges and any deposits not yet recorded by the bank. This is the checkbook balance.

4. Compare the balance with the bank balance listed on the bank statement. If the two numbers are equal, the bank statement and checkbook balance.

Current checkbook balance:	1113.90
Checks: 265	67.14
267	43.85
271	7.00
Interest:	+ 4.47
	1236.36
Service charge:	− 3.00
	1233.36
Deposit:	− 225.00
Checkbook balance:	1008.36

Closing bank balance   Checkbook
from bank statement   balance

$1008.36   = $1008.36

The bank statement and checkbook balance.

RECORD ALL CHARGES OR CREDITS THAT AFFECT YOUR ACCOUNT								
NUMBER	DATE	DESCRIPTION OF TRANSACTION	PAYMENT/DEBIT (-)	√T	FEE (IF ANY) (-)	DEPOSIT/CREDIT (+)	BALANCE $ 1620	42
413	3/2	Car Payment	$132 15	√	$	$	1488	27
414	3/2	Utility	67 14	√			1421	13
415	3/5	Restaurant - Dinner for 4	78 14				1342	99
	3/8	Deposit		√		1842 66	3185	65
416	3/10	House Payment	672 14	√			2513	51
417	3/14	Insurance	177 10				2336	41

CHECKING ACCOUNT Monthly Statement   Account Number: 924-297-8

Date	Transaction	Amount	Balance
3/1	OPENING BALANCE		1620.42
3/4	CHECK	132.15	1488.27
3/5	CHECK	67.14	1421.13
3/8	DEPOSIT	1842.66	3263.79
3/10	INTEREST	6.77	3270.56
3/12	CHECK	672.14	2598.42
3/25	SERVICE CHARGE	2.00	2596.42
3/30	CLOSING BALANCE		2596.42

Balance the bank statement shown above.

1. In the checkbook register, put a check (√) by each check paid by the bank and each deposit recorded by the bank.

2. Add to the current checkbook balance all checks that have been written but not yet paid by the bank and any interest paid on the account.

3. Subtract any service charges and any deposits not yet recorded by the bank. This is the checkbook balance.

4. Compare the balance with the bank balance listed on the bank statement. If the two numbers are equal, the bank statement and checkbook balance.

```
Current checkbook
balance:                2336.41
Checks:  415              78.14
         417             177.10
Interest:              +   6.77
                        2598.42
Service charge:        -   2.00
                        2596.42

Checkbook Balance:      2596.42
```

Closing bank balance   Checkbook
from bank statement   balance:

    $2596.42   = $2596.42

The bank statement and checkbook balance.

		RECORD ALL CHARGES OR CREDITS THAT AFFECT YOUR ACCOUNT						BALANCE	
NUMBER	DATE	DESCRIPTION OF TRANSACTION	PAYMENT/DEBIT (-)	√T	FEE (IF ANY) (-)	DEPOSIT/CREDIT (+)		$ 412	64
345	1/14	Phone bill	$ 34 75		$	$		377	89
346	1/19	magazine	8 98					368	91
347	1/23	Theatre tickets	45 00					323	91
	1/31	Deposit				947	00	1270	91
348	2/5	Cash	250 00					1020	91
349	2/12	Rent	440 00					580	91

CHECKING ACCOUNT Monthly Statement   Account Number: 924-297-8

Date	Transaction	Amount	Balance
1/10	OPENING BALANCE		412.64
1/18	CHECK	34.75	377.89
1/23	CHECK	8.98	368.91
1/31	DEPOSIT	947.00	1315.91
2/1	INTEREST	4.52	1320.43
2/10	CHECK	250.00	1070.43
2/10	CLOSING BALANCE		1070.43

Example 3

Balance the bank statement shown above.

Solution

Current checkbook
balance: 580.91
Checks: 347   45.00
        349   440.00
Interest: + 4.52
                       1070.43
Service charge: −   0.00
                       1070.43
Deposit: −   0.00
Checkbook balance: 1070.43

Closing bank balance from bank statement: $1070.43

Checkbook balance: $1070.43

The bank statement and checkbook balance.

		RECORD ALL CHARGES OR CREDITS THAT AFFECT YOUR ACCOUNT						
NUMBER	DATE	DESCRIPTION OF TRANSACTION	PAYMENT/DEBIT (-)	√T	FEE (IF ANY) (-)	DEPOSIT/CREDIT (+)	BALANCE $ 603	17
	2/15	Deposit	$		$	$ 523 84	1127	01
234	2/20	Mortgage	473 21				653	80
235	2/27	Cash	200 00				453	80
	3/1	Deposit				523 84	977	64
236	3/12	Insurance	275 50				702	14
237	3/12	Telephone	48 73				653	41

CHECKING ACCOUNT Monthly Statement			Account Number: 314-271-4
Date	Transaction	Amount	Balance
2/14	OPENING BALANCE		603.17
2/15	DEPOSIT	523.84	1127.01
2/21	CHECK	473.21	653.80
2/28	CHECK	200.00	453.80
3/1	INTEREST	2.11	455.91
3/14	CHECK	275.50	180.41
3/14	CLOSING BALANCE		180.41

Example 4

Balance the bank statement shown above.

Your solution

Solution on p. A28

6.7 EXERCISES

▶ Objective A *Application Problems*

1. You had a checking account balance of $342.51 before making a deposit of $143.81. What is your new checking account balance?

2. A student had a checking account balance of $493.26 before writing a check for $48.39. What is the current checking account balance?

3. A real estate firm had a balance of $2431.76 in its rental property checking account. What is the balance in this account after a check for $1209.29 has been written?

4. The business checking account for a tire store showed a balance of $1536.97. What is the balance in this account after a deposit of $439.21 has been made?

5. A nutritionist had a checking account balance of $1204.63 before writing one check for $119.27 and another check for $260.09. Find the current checkbook balance.

6. A farmer had a checking account balance of $3046.93 before writing a check for $1027.33 and making a deposit of $150.00. Find the current checkbook balance.

7. The business checking account for a dry cleaning store had a balance of $3476.85 before a deposit of $1048.53 was made. The store manager then wrote checks, one for $848.37 and another for $676.19. Find the current checkbook balance.

8. A pottery store manager had a checking account balance of $427.38 before a deposit of $127.29 was made. The manager then wrote two checks, one for $43.52 and one for $249.78. Find the current checkbook balance.

9. A carpenter had a checkbook balance of $104.96 before making a deposit of $350 and writing a check for $71.29. Is there enough money in the account to purchase a refrigerator for $375?

10. A taxidriver had a checkbook balance of $149.85 before making a deposit of $245 and writing a check for $387.68. Is there enough money in the account for the bank to pay the check?

11. A sporting goods store has the opportunity to buy downhill skis and cross-country skis at a manufacturer's close-out sale. The downhill skis will cost $3500, and the cross-country skis will cost $2050. There is currently $5625.42 in the sporting goods store checking account. Is there enough money in the account to make both purchases by check?

12. A lathe operator's current checkbook balance is $643.42. The operator wants to purchase a utility trailer for $225 and a used piano for $450. Is there enough money in the account to make the two purchases?

Objective B Application Problems

13. Balance the checkbook.

NUMBER	DATE	DESCRIPTION OF TRANSACTION	PAYMENT/DEBIT (-)	√T	FEE (IF ANY) (-)	DEPOSIT/CREDIT (+)	BALANCE $ 1035 18	
218	7/2	Mortgage	$284 60		$	$	750	58
219	7/4	Telephone	23 36				727	22
220	7/7	Cash	200 00				527	22
	7/12	Deposit				792 60	1319	82
221	7/15	Insurance	192 30				1127	52
222	7/18	Investment	100 00				1027	52
223	7/20	Credit card	214 83				812	69
	7/26	Deposit				792 60	1605	29
224	7/27	Department store	113 37				1491	92

CHECKING ACCOUNT Monthly Statement   Account Number: 122-345-1

Date	Transaction	Amount	Balance
7/1	OPENING BALANCE		1035.18
7/1	INTEREST	5.15	1040.33
7/4	CHECK	284.60	755.73
7/6	CHECK	23.36	732.37
7/12	DEPOSIT	792.60	1524.97
7/20	CHECK	192.30	1332.67
7/24	CHECK	100.00	1232.67
7/26	DEPOSIT	792.60	2025.27
7/28	CHECK	200.00	1825.27
7/30	CLOSING BALANCE		1825.27

14. Balance the checkbook.

NUMBER	DATE	DESCRIPTION OF TRANSACTION	PAYMENT/DEBIT (-)	√T	FEE (IF ANY) (-)	DEPOSIT/CREDIT (+)	BALANCE $ 466.79	
223	3/2	Groceries	$ 67.32		$	$	399	47
	3/5	Deposit				560.70	960	17
224	3/5	Rent	460.00				500	17
225	3/7	Gas & electric	42.35				457	82
226	3/7	Cash	100.00				357	82
227	3/7	Insurance	118.44				239	38
228	3/7	Credit card	119.32				120	06
229	3/12	Dentist	42.00				78	06
230	3/13	Drug store	17.03				61	03
	3/19	Deposit				560.70	621	73
231	3/22	Car payment	141.35				480	38
232	3/25	Cash	100.00				380	38
233	3/25	Oil company	66.40				313	98
234	3/28	Plumber	55.73				258	25
235	3/29	Department store	88.39				169	86
	3/30							

CHECKING ACCOUNT Monthly Statement		Account Number: 122-345-1	
Date	Transaction	Amount	Balance
3/1	OPENING BALANCE		466.79
3/5	DEPOSIT	560.70	1027.49
3/7	CHECK	67.32	960.17
3/8	CHECK	460.00	500.17
3/8	CHECK	100.00	400.17
3/9	CHECK	42.35	357.82
3/12	CHECK	118.44	239.38
3/14	CHECK	42.00	197.38
3/18	CHECK	17.03	180.35
3/19	DEPOSIT	560.70	741.05
3/25	CHECK	141.35	599.70
3/27	CHECK	100.00	499.70
3/29	CHECK	55.73	443.97
3/30	INTEREST	13.22	457.19
4/1	CLOSING BALANCE		457.19

15. Balance the checkbook.

NUMBER	DATE	DESCRIPTION OF TRANSACTION	PAYMENT/DEBIT (-)	√T	FEE (IF ANY) (-)	DEPOSIT/CREDIT (+)	BALANCE $ 219 43
	5/1	Deposit				219 14	438 57
515	5/2	Electric Bill	22 35				416 22
516	5/2	Groceries	55 14				361 08
517	5/4	Insurance	122 17				238 91
518	5/5	Theatre tickets	24 50				214 41
	5/8	Deposit				219 14	433 55
519	5/10	Telephone	17 39				416 16
520	5/12	Newspaper	12 50				403 66
	5/15	Interest				7 82	411 48
	5/15	Deposit				219 14	630 62
521	5/20	Hotel	172 90				457 72
522	5/21	Credit card	113 44				344 28
523	5/23	Eye Exam	42 00				302 28
524	5/24	Groceries	77 14				225 14
525	5/24	Deposit				219 14	444 28
526	5/25	Oil company	44 16				400 12
527	5/30	Car payment	88 62				311 50
528	5/30	Doctor	37 42				274 08

CHECKING ACCOUNT Monthly Statement   Account Number: 122-345-1

Date	Transaction	Amount	Balance
5/1	OPENING BALANCE		219.43
5/1	DEPOSIT	219.14	438.57
5/3	CHECK	55.14	383.43
5/4	CHECK	22.35	361.08
5/6	CHECK	24.50	336.58
5/8	CHECK	122.17	214.41
5/8	DEPOSIT	219.14	433.55
5/15	INTEREST	7.82	441.37
5/15	CHECK	17.39	423.98
5/15	DEPOSIT	219.14	643.12
5/23	CHECK	42.00	601.12
5/23	CHECK	172.90	428.22
5/24	CHECK	77.14	351.08
5/24	DEPOSIT	219.14	570.22
5/30	CHECK	88.62	481.60
6/1	CLOSING BALANCE		481.60

Calculators and Computers

Amortization Did you ever wonder how your car payments are calculated? Maybe you own a house and make a monthly mortgage payment. In both these cases the amount of your payment is calculated so that you repay the amount you borrowed (the principal) and the interest you were charged to borrow the money. In the words of a banker, your monthly payments "amortize" the loan.

The series of monthly payments you make is called an amortization schedule. The program AMORTIZATION on the Math ACE disk can be used to calculate your monthly payment on any fixed interest loan. The program will also tell you the amount of interest you will repay.

Here are three guidelines that will make using the program a little easier:

1. Enter the annual interest rate as a percent in decimal form. For example, if the interest rate is $9\frac{1}{4}\%$, enter the interest rate as 9.25.
2. Because the payments are *monthly*, you must enter the number of months of the loan. For example, if the loan is for 5 years, then the number of months is 60 (5 × 12).
3. Enter the amount of money borrowed. Do not use dollar signs ($) or commas in your number.

Chapter Summary

Key Words The *unit cost* is the cost of one item.

Percent increase is used to show how much a quantity has increased over its original value.

Cost is the price that a business pays for a product.

Selling price is the price at which a business sells a product to a customer.

Markup is the difference between selling price and cost.

Markup rate is frequently expressed as a percent of a product's cost.

Percent decrease is used to show how much a quantity has decreased from its original value.

Sale price is the reduced price from the regular price.

Discount is the difference between the regular price and the sale price.

Discount rate is frequently stated as a percent of a product's regular price.

Interest is the amount of money paid for the privilege of using someone else's money.

Principal is the amount of money originally deposited or borrowed.

The percent used to determine the amount of interest is the *interest rate*.

Interest computed on the original amount is called *simple interest*.

Compound interest is computed not only on the original principal but also on interest already earned.

The *mortgage* is the amount that is borrowed to buy real estate.

The loan origination fee is usually a percent of the mortgage and is expressed in *points*.

Commissions are usually paid to salespersons and are calculated as a percent of total sales.

An employee who receives an *hourly wage* is paid a certain amount for each hour worked.

An employee who is paid a *salary* receives payment based on a weekly, biweekly, monthly, or annual time schedule.

Essential Rules

To Find Unit Cost   To find the unit cost, divide the total cost by the number of units.

To Find Total Cost   To find the total cost, multiply the unit cost by the number of units.

Basic Markup Equations:

Selling Price = cost + markup
$$S = C + M$$

Markup = markup rate × cost
$$M = r \times C$$

Basic Discount Equations:

Sale Price = regular price − discount
$$S = R - D$$

Discount = discount rate × regular price
$$D = r \times R$$

Annual Simple Interest Equation:

$$\text{Principal} \times \text{annual interest rate} \times \text{time in years} = \text{Interest}$$
$$P \times r \times t = I$$

Chapter Review

SECTION 6.1

1. A 20-ounce box of cereal costs $2.90. Find the unit cost.

2. Twenty-four ounces of a mouthwash cost $3.49. A 60-ounce container of the same kind of mouthwash costs $8.40. Which is the better buy?

SECTION 6.2

3. An oil stock was bought for 42\frac{3}{8}$ per share; 6 months later the stock was selling for 55\frac{1}{4}$ per share. Find the percent increase in the price of the stock for 6 months. Round to the nearest tenth of a percent.

4. A professional baseball player received a salary of $1 million last year. This year the player signed a contract paying $12 million over 4 years. Find the yearly percent increase in the player's salary.

5. A computer store uses a markup rate of 35% on all computer systems. Find the selling price of a computer system that costs the store $1540.

6. Last year an oil company had earnings of $4.12 per share. This year the earnings are $4.73 per share. What is the percent increase in earnings per share? Round to the nearest percent.

7. A sporting goods store uses a markup rate of 40%. What is the markup on a ski suit that costs the store $180?

8. A suit that regularly costs $235 is on sale for 40% off the regular price. Find the sale price.

SECTION 6.3

9. A contractor borrows $100,000 from a credit union for 9 months at an annual interest rate of 11%. What is the simple interest due on the loan?

10. A sporting goods store borrowed $30,000 at an annual interest rate of 11% for 6 months. Find the simple interest due on the loan.

11. A computer programmer invests $25,000 in a retirement account that pays 9% interest, compounded daily. What is the value of the investment in 10 years? Use the table on page A4. Round to the nearest cent.

12. A fast food restaurant invests $50,000 in an account that pays 11% annual interest, compounded quarterly. What will be the value of the investment in 1 year? Use the table on page A4.

SECTION 6.4

13. A home buyer purchases a home for $125,000. The lender requires a down payment of 15%. Find the amount of the down payment.

14. A credit union requires a borrower to pay $2\frac{1}{2}$ points for a loan. Find the origination fee for a loan of $75,000.

15. The monthly mortgage payment for a condominium is $523.67. The owner must pay an annual property tax of $658.32. Find the total monthly payment for the mortgage and property tax.

16. A family bought a home for $156,000. The family made a 10% down payment and financed the remainder with a 30-year loan with an annual interest rate of 10%. Find the monthly mortgage payment. Use the monthly payment table on page A6. Round to the nearest cent.

SECTION 6.5

17. A plumber buys a truck for $13,500. A state license of $315 and a sales tax of 6.25% of the purchase price are required. Find the total cost of the sales tax and the license fee.

18. An account executive spent $1025.58 for insurance, $605.82 for gas, $37.92 for oil, and $188.27 for maintenance during a year in which 15,320 miles were driven. Find the cost per mile for these four items. Round to the nearest tenth of a cent.

19. A car owner pays a monthly car payment of $122.78. During a month when $25.45 is principal, how much of the payment is interest?

20. A pickup truck with a slide-in camper is purchased for $14,450. A down payment of 8% is made, and the remaining cost is financed for 4 years at an annual interest rate of 11%. Find the monthly payment. Use the monthly payment schedule on page A6. Round to the nearest cent.

SECTION 6.6

21. The manager of the retail store at a ski resort receives a commission of 3% on all sales at the alpine shop. Find the total commission received during a month in which the shop had $108,000 of sales.

22. A nurse receives $12.60 per hour for working 40 hours a week and time and a half for working over 40 hours. Find the nurse's total income during a week in which the nurse worked 48 hours.

SECTION 6.7

23. A student had a checking account balance of $1568.45 before writing checks for $123.76, $756.45, and $88.77. The student then deposited a check for $344.21. Find the student's current checkbook balance.

24. The business checking account of a donut shop showed a balance of $9567.44 before checks of $1023.55, $345.44, and $23.67 were written and checks of $555.89 and $135.91 were deposited. Find the current checkbook balance.

Chapter Test

1. Twenty feet of lumber cost $138.40. What is the cost per foot?

2. Find the more economical purchase: 5 pounds of tomatoes for $1.65, or 8 pounds for $2.72.

3. Red snapper cost $4.15 per pound. Find the cost of $3\frac{1}{2}$ pounds. (Round to the nearest cent.)

4. An exercise bicycle increased in price from $415 to $498. Find the percent increase in the cost of the exercise bicycle.

5. Fifteen years ago a painting was priced at $6000. Today the same painting has a value of $15,000. Find the percent increase in the price of the painting during the fifteen years.

6. A department store uses a 40% markup rate. Find the selling price of a compact disc player that the store purchased for $215.

7. A bookstore buys a paperback book for $5 and uses a markup rate of 25%. Find the selling price of the book.

8. The price of gold dropped from $850 per ounce to $360 per ounce. What percent decrease does this amount represent? (Round to the nearest tenth of a percent.)

9. The price of a video camera drops from $1120 to $896. What percent decrease does this price drop represent?

10. A corner hutch with a regular price of $299 is on sale for 30% off the regular price. Find the sale price.

11. A box of stationery that regularly sells for $4.50 is on sale for $2.70. Find the discount rate.

12. A construction company borrowed $75,000 at an annual interest rate of 11% for 4 months. Find the simple interest due on the loan.

13. A self-employed individual placed $30,000 in an account that pays 9% annual interest, compounded quarterly. How much interest was earned in 10 years? Use the table on page A4.

14. A savings and loan institution is giving mortgage loans that have a loan origination fee of $2\frac{1}{2}$ points. Find the loan origination fee on a home purchased with a loan of $134,000.

15. A new housing development offers homes with a mortgage of $122,000 for 25 years at an annual interest rate of 10%. Find the monthly mortgage payment. Use the table on page A6.

16. A $17,500 minivan is purchased with an 18% down payment. Find the amount financed.

17. A rancher buys a pickup for $11,500, with a down payment of 15% of the cost. The balance is financed for 3 years at an annual interest rate of 11%. Find the monthly car payment. Use the table on page A6.

18. An emergency room nurse receives an hourly wage of $13.40 an hour. When called in at night, the nurse receives time and a half. How much does the nurse earn in a week when working 30 hours at normal rates and 15 hours during the night?

19. The business checking account for a pottery store had a balance of $7349.44 before checks for $1349.67 and $344.12 were written. The store manager then made a deposit of $956.60. Find the current checkbook balance.

20. Balance the checkbook shown.

RECORD ALL CHARGES OR CREDITS THAT AFFECT YOUR ACCOUNT

NUMBER	DATE	DESCRIPTION OF TRANSACTION	PAYMENT/DEBIT (-)	√T	FEE (IF ANY) (-)	DEPOSIT/CREDIT (+)	BALANCE $ 422 13
	8/1	House Payment	$ 213 72		$	$	208 41
	8/4	Deposit				552 60	761 01
	8/5	Plane Tickets	162 40				598 61
	8/6	Groceries	66 44				532 17
x	8/10	Car Payment	122 37				409 80
	8/15	Deposit				552 60	962 40
x	8/16	Credit card	213 45				748 95
	8/18	Doctor	92 14				656 81
	8/22	Utilities	72 30				584 51
x	8/28	T.V. Repair	78 20				506 31

CHECKING ACCOUNT Monthly Statement		Account Number: 122-345-1	
Date	Transaction	Amount	Balance
8/1	OPENING BALANCE		422.13
8/3	CHECK	213.72	208.41
8/4	DEPOSIT	552.60	761.01
8/8	CHECK	66.44	694.57
8/8	CHECK	162.40	532.17
8/15	DEPOSIT	552.60	1084.77
8/23	CHECK	72.30	1012.47
8/24	CHECK	92.14	920.33
9/1	CLOSING BALANCE		920.33

Cumulative Review

1. Simplify $12 - (10 - 8)^2 \div 2 + 3$.

2. Add: $3\frac{1}{3} + 4\frac{1}{8} + 1\frac{1}{12}$

3. Subtract: $12\frac{3}{16} - 9\frac{5}{12}$

4. Multiply: $5\frac{5}{8} \times 1\frac{9}{15}$

5. Divide: $3\frac{1}{2} \div 1\frac{3}{4}$

6. Simplify $\left(\frac{3}{4}\right)^2 \div \left(\frac{3}{8} - \frac{1}{4}\right) + \frac{1}{2}$.

7. Divide: $0.059\overline{)3.0792}$. Round to the nearest tenth.

8. Convert $\frac{17}{12}$ to a decimal. Round to the nearest thousandth.

9. Write "$410 in 8 hours" as a unit rate.

10. Solve the proportion $\frac{5}{n} = \frac{16}{35}$. Round to the nearest hundredth.

11. Write $\frac{5}{8}$ as a percent.

12. Find 6.5% of 420.

13. Write 18.2% as a decimal.

14. What percent of 20 is 8.4?

15. 30 is 12% of what?

16. 65 is 42% of what? Round to the nearest hundredth.

Cumulative Review

17. A series of late summer storms produced rainfall of $3\frac{3}{4}$, $8\frac{1}{2}$, and $1\frac{2}{3}$ inches during a 3-week period. Find the total rainfall during the 3 weeks.

18. A family pays $\frac{1}{5}$ of its total monthly income for taxes. The family has a total monthly income of $2850. Find the amount of the monthly income that the family pays in taxes.

19. In 5 years, the cost of a scientific calculator went from $75 to $30. What is the ratio of the decrease in price to the original price?

20. A compact car drove 417.5 miles on 12.5 gallons of gasoline. Find the number of miles driven per gallon of gasoline.

21. A 14 pound turkey costs $12.96. Find the unit cost. (Round to the nearest cent.)

22. Eighty shares of a stock has a dividend of $112. At the same rate, find the dividend on 200 shares of the stock.

23. A video camera that regularly sells for $900 is on sale for 20% off the regular price. What is the sale price?

24. A department store buys a portable disc player for $85 and uses a markup rate of 40%. Find the selling price of the disc player.

25. An elementary school teacher received an increase in salary from $2800 per month to $3024 per month. Find the percent increase in the teacher's salary.

26. A contractor borrowed $120,000 for 6 months at an annual interest rate of 10%. How much simple interest is due on the loan?

27. A red sports car is purchased for $14,000, and a down payment of $2000 is made. The balance is financed for 3 years at an annual interest rate of 9%. Find the monthly payment. Use the table on page A6. (Round to the nearest cent.)

28. A family had a checking account balance of $1846.78. A check of $568.30 was deposited into the account, and checks of $123.98 and $47.33 were written. Find the new checking account balance.

29. A car owner spent $840 on gasoline and oil, $520 on insurance, $185 on tires, and $432 on repairs. Find the cost per mile to drive the car 10,000 miles during the year. (Round to the nearest cent.)

30. A house has a mortgage of $72,000 for 20 years at an annual interest rate of 11%. Find the monthly mortgage payment. Use the table on page A6. (Round to the nearest cent.)

APPENDIX

Addition Table

+	0	1	2	3	4	5	6	7	8	9	10	11	12
0	0	1	2	3	4	5	6	7	8	9	10	11	12
1	1	2	3	4	5	6	7	8	9	10	11	12	13
2	2	3	4	5	6	7	8	9	10	11	12	13	14
3	3	4	5	6	7	8	9	10	11	12	13	14	15
4	4	5	6	7	8	9	10	11	12	13	14	15	16
5	5	6	7	8	9	10	11	12	13	14	15	16	17
6	6	7	8	9	10	11	12	13	14	15	16	17	18
7	7	8	9	10	11	12	13	14	15	16	17	18	19
8	8	9	10	11	12	13	14	15	16	17	18	19	20
9	9	10	11	12	13	14	15	16	17	18	19	20	21
10	10	11	12	13	14	15	16	17	18	19	20	21	22
11	11	12	13	14	15	16	17	18	19	20	21	22	23
12	12	13	14	15	16	17	18	19	20	21	22	23	24

Multiplication Table

×	0	1	2	3	4	5	6	7	8	9	10	11	12
0	0	0	0	0	0	0	0	0	0	0	0	0	0
1	0	1	2	3	4	5	6	7	8	9	10	11	12
2	0	2	4	6	8	10	12	14	16	18	20	22	24
3	0	3	6	9	12	15	18	21	24	27	30	33	36
4	0	4	8	12	16	20	24	28	32	36	40	44	48
5	0	5	10	15	20	25	30	35	40	45	50	55	60
6	0	6	12	18	24	30	36	42	48	54	60	66	72
7	0	7	14	21	28	35	42	49	56	63	70	77	84
8	0	8	16	24	32	40	48	56	64	72	80	88	96
9	0	9	18	27	36	45	54	63	72	81	90	99	108
10	0	10	20	30	40	50	60	70	80	90	100	110	120
11	0	11	22	33	44	55	66	77	88	99	110	121	132
12	0	12	24	36	48	60	72	84	96	108	120	132	144

Table of Square Roots

Decimal approximations have been rounded to the nearest thousandth.

Number	Square Root	Number	Square Root	Number	Square Root	Number	Square Root
1	1	51	7.141	101	10.050	151	12.288
2	1.414	52	7.211	102	10.100	152	12.329
3	1.732	53	7.280	103	10.149	153	12.369
4	2	54	7.348	104	10.198	154	12.410
5	2.236	55	7.416	105	10.247	155	12.450
6	2.449	56	7.483	106	10.296	156	12.490
7	2.646	57	7.550	107	10.344	157	12.530
8	2.828	58	7.616	108	10.392	158	12.570
9	3	59	7.681	109	10.440	159	12.610
10	3.162	60	7.746	110	10.488	160	12.649
11	3.317	61	7.810	111	10.536	161	12.689
12	3.464	62	7.874	112	10.583	162	12.728
13	3.606	63	7.937	113	10.630	163	12.767
14	3.742	64	8	114	10.677	164	12.806
15	3.873	65	8.062	115	10.724	165	12.845
16	4	66	8.124	116	10.770	166	12.884
17	4.123	67	8.185	117	10.817	167	12.923
18	4.243	68	8.246	118	10.863	168	12.961
19	4.359	69	8.307	119	10.909	169	13
20	4.472	70	8.367	120	10.954	170	13.038
21	4.583	71	8.426	121	11	171	13.077
22	4.690	72	8.485	122	11.045	172	13.115
23	4.796	73	8.544	123	11.091	173	13.153
24	4.899	74	8.602	124	11.136	174	13.191
25	5	75	8.660	125	11.180	175	13.229
26	5.099	76	8.718	126	11.225	176	13.267
27	5.196	77	8.775	127	11.269	177	13.304
28	5.292	78	8.832	128	11.314	178	13.342
29	5.385	79	8.888	129	11.358	179	13.379
30	5.477	80	8.944	130	11.402	180	13.416
31	5.568	81	9	131	11.446	181	13.454
32	5.657	82	9.055	132	11.489	182	13.491
33	5.745	83	9.110	133	11.533	183	13.528
34	5.831	84	9.165	134	11.576	184	13.565
35	5.916	85	9.220	135	11.619	185	13.601
36	6	86	9.274	136	11.662	186	13.638
37	6.083	87	9.327	137	11.705	187	13.675
38	6.164	88	9.381	138	11.747	188	13.711
39	6.245	89	9.434	139	11.790	189	13.748
40	6.325	90	9.487	140	11.832	190	13.784
41	6.403	91	9.539	141	11.874	191	13.820
42	6.481	92	9.592	142	11.916	192	13.856
43	6.557	93	9.644	143	11.958	193	13.892
44	6.633	94	9.695	144	12	194	13.928
45	6.708	95	9.747	145	12.042	195	13.964
46	6.782	96	9.798	146	12.083	196	14
47	6.856	97	9.849	147	12.124	197	14.036
48	6.928	98	9.899	148	12.166	198	14.071
49	7	99	9.950	149	12.207	199	14.107
50	7.071	100	10	150	12.247	200	14.142

Compound Interest Table

Compounded Annually

	5%	6%	7%	8%	9%	10%	11%
1 year	1.05000	1.06000	1.07000	1.08000	1.09000	1.10000	1.11000
5 years	1.27628	1.33823	1.40255	1.46933	1.53862	1.61051	1.68506
10 years	1.62890	1.79085	1.96715	2.15893	2.36736	2.59374	2.83942
15 years	2.07893	2.39656	2.75903	3.17217	3.64248	4.17725	4.78459
20 years	2.65330	3.20714	3.86968	4.66095	5.60441	6.72750	8.06239

Compounded Semiannually

	5%	6%	7%	8%	9%	10%	11%
1 year	1.05062	1.06090	1.07123	1.08160	1.09203	1.10250	1.11303
5 years	1.28008	1.34392	1.41060	1.48024	1.55297	1.62890	1.70814
10 years	1.63862	1.80611	1.98979	2.19112	2.41171	2.65330	2.91776
15 years	2.09757	2.42726	2.80679	3.24340	3.74531	4.32194	4.98395
20 years	2.68506	3.26204	3.95926	4.80102	5.81634	7.03999	8.51331

Compounded Quarterly

	5%	6%	7%	8%	9%	10%	11%
1 year	1.05094	1.06136	1.07186	1.08243	1.09308	1.10381	1.11462
5 years	1.28204	1.34686	1.41478	1.48595	1.56051	1.63862	1.72043
10 years	1.64362	1.81402	2.00160	2.20804	2.43519	2.68506	2.95987
15 years	2.10718	2.44322	2.83182	3.28103	3.80013	4.39979	5.09225
20 years	2.70148	3.29066	4.00639	4.87544	5.93015	7.20957	8.76085

Compounded Daily

	5%	6%	7%	8%	9%	10%	11%
1 year	1.05127	1.06183	1.07250	1.08328	1.09416	1.10516	1.11626
5 years	1.28400	1.34983	1.41902	1.49176	1.56823	1.64861	1.73311
10 years	1.64866	1.82203	2.01362	2.22535	2.45933	2.71791	3.00367
15 years	2.11689	2.45942	2.85736	3.31968	3.85678	4.48077	5.20569
20 years	2.71810	3.31979	4.05466	4.95217	6.04830	7.38703	9.02203

To use this table:
1. Locate the section which gives the desired compounding period.
2. Locate the interest rate in the top row of that section.
3. Locate the number of years in the left-hand column of that section.
4. Locate the number where the interest-rate column and the number-of-years row meet. This is the compound interest factor.

Example An investment yields an annual interest rate of 10% compounded quarterly for 5 years.
The compounding period is "compounded quarterly."
The interest rate is 10%.
The number of years is 5.
The number where the row and column meet is 1.63862. This is the compound interest factor.

Compound Interest Table

Compounded Annually

	12%	13%	14%	15%	16%	17%	18%
1 year	1.12000	1.13000	1.14000	1.15000	1.16000	1.17000	1.18000
5 years	1.76234	1.84244	1.92542	2.01136	2.10034	2.19245	2.28776
10 years	3.10585	3.39457	3.70722	4.04556	4.41144	4.80683	5.23384
15 years	5.47357	6.25427	7.13794	8.13706	9.26552	10.53872	11.97375
20 years	9.64629	11.52309	13.74349	16.36654	19.46076	23.10560	27.39303

Compounded Semiannually

	12%	13%	14%	15%	16%	17%	18%
1 year	1.12360	1.13423	1.14490	1.15563	1.16640	1.17723	1.18810
5 years	1.79085	1.87714	1.96715	2.06103	2.15893	2.26098	2.36736
10 years	3.20714	3.52365	3.86968	4.24785	4.66096	5.11205	5.60441
15 years	5.74349	6.61437	7.61226	8.75496	10.06266	11.55825	13.26768
20 years	10.28572	12.41607	14.97446	18.04424	21.72452	26.13302	31.40942

Compounded Quarterly

	12%	13%	14%	15%	16%	17%	18%
1 year	1.12551	1.13648	1.14752	1.15865	1.16986	1.18115	1.19252
5 years	1.80611	1.89584	1.98979	2.08815	2.19112	2.29891	2.41171
10 years	3.26204	3.59420	3.95926	4.36038	4.80102	5.28497	5.81636
15 years	5.89160	6.81402	7.87809	9.10513	10.51963	12.14965	14.02741
20 years	10.64089	12.91828	15.67574	19.01290	23.04980	27.93091	33.83010

Compounded Daily

	12%	13%	14%	15%	16%	17%	18%
1 year	1.12747	1.13880	1.15024	1.16180	1.17347	1.18526	1.19716
5 years	1.82194	1.91532	2.01348	2.11667	2.22515	2.33918	2.45906
10 years	3.31946	3.66845	4.05411	4.48031	4.95130	5.47178	6.04696
15 years	6.04786	7.02625	8.16288	9.48335	11.01738	12.79950	14.86983
20 years	11.01883	13.45751	16.43582	20.07316	24.51534	29.94039	36.56578

Monthly Payment Table

	8%	9%	10%	11%	12%	13%
1 year	0.0869884	0.0874515	0.0879159	0.0883817	0.0888488	0.0893173
2 years	0.0452273	0.0456847	0.0461449	0.0466078	0.0470735	0.0475418
3 years	0.0313364	0.0317997	0.0322672	0.0327387	0.0332143	0.0336940
4 years	0.0244129	0.0248850	0.0253626	0.0258455	0.0263338	0.0268275
5 years	0.0202764	0.0207584	0.0212470	0.0217424	0.0222445	0.0227531
20 years	0.0083644	0.0089973	0.0096502	0.0103219	0.0110109	0.0117158
25 years	0.0077182	0.0083920	0.0090870	0.0098011	0.0105322	0.0112784
30 years	0.0073376	0.0080462	0.0087757	0.0095232	0.0102861	0.0110620

To use this table:
1. Locate the desired interest rate in the top row.
2. Locate the number of years in the left-hand column.
3. Locate the number where the interest-rate column and the number-of-years row meet. This is the monthly payment factor.

Example: A home has a 30-year mortgage at an annual interest rate of 12%.
The interest rate is 12%.
The number of years is 30.
The number where the row and column meet is 0.0102861. This is the monthly payment factor.

SOLUTIONS to Chapter 1 Examples

SECTION 1.1 *pages 3–6*

Example 2 (number line from 0 to 12 with point at 9)

Example 6 Thirty-six million four hundred sixty-two thousand seventy-five

Example 10 60,000 + 8000 + 200 + 80 + 1

Example 14 370,000

Example 4 a. 45 > 29   b. 27 > 0

Example 8 452,007

Example 12 100,000 + 9000 + 200 + 7

Example 16 4000

SECTION 1.2 *pages 9–12*

Example 2
```
   3,508
 +92,170
 ──────
  95,678
```

Example 6
```
   2
   95
   88
  +67
  ───
  250
```

Example 4
```
   102
  7351
  1024
 + 410
 ─────
  8887
```

Example 8
```
  1 1 2 1
     392
   4,079
  89,035
 + 4,992
 ───────
  98,498
```

Example 10

Strategy To find the odometer reading after the trip, add the mileage before the trip (23,972) to the miles traveled during the trip (297).

Solution
```
  23,972
 +   297
 ───────
  24,269
```

The odometer reading after the trip was 24,269 miles.

Example 12

Strategy To find the total amount budgeted for the three items each month, add the three amounts ($275, $75, and $64).

Solution
```
  $275
    75
  + 64
  ────
  $414
```

The total amount budgeted for the three items is $414.

SECTION 1.3 *pages 19–22*

Example 2
```
  8925
 -6413
  2512
```

Example 4
```
 17,504
- 9,302
  8,202
```

Example 6
```
  2 14 7 11
  3̸ 4̸ 8̸ 1̸
-    8 6 5
    2 6 1 6
```

Example 8
```
          15
        4 6 12
  5 4, 5̸ 6̸ 2̸
 -1 4, 4 8 5
  4 0, 0 7 7
```

Example 10
```
         13 9 9
     5 3 10 10 13
     6̸ 4̸, 0̸ 0̸ 3̸
    -5 4, 9 3 6
     9, 0 6 7
```

Example 12

Strategy To find the amount that remains to be paid, subtract the down payment ($675) from the cost ($3250).

Solution
```
       11 14
     2 1̸ 4̸ 10
   $ 3̸ 2̸ 5̸ 0̸
   -   6 7 5
   $ 2 5 7 5
```
$2575 remains to be paid.

Example 14

Strategy To find your take-home pay:
- Add to find the total of the deductions ($127 + $18 + $35).
- Subtract the total of the deductions from your total salary ($638).

Solution
```
  $127            $638
    18           -180
  + 35            $458
  $180 deductions
```

Your take-home pay is $458.

SECTION 1.4 *pages 27–30*

Example 2
```
    4
   78
 ×  6
  468
```

Example 4
```
    3 5
   648
 ×   7
  4536
```

Example 6
```
     756
   ×305
    3780
   22680
  230,580
```

Example 8

Strategy To find the number of cars the dealer will receive in 12 months, multiply the number of months (12) by the number of cars received each month (37).

Solution
```
   37
 ×12
   74
  37
  444
```

The dealer will receive 444 cars in 12 months.

Example 10

Strategy To find the total cost of the order:
- Find the cost of the sports jackets by multiplying the number of jackets (25) by the cost for each jacket ($23).
- Add the product to the cost for the suits ($4800).

Solution
```
  $23            $4800
  ×25          +  575
  115           $5375
  46
 $575   cost for jackets
```

The total cost of the order is $5375.

SECTION 1.5 pages 35–42

Example 2
$$\begin{array}{r}7\\9\overline{)63}\end{array}$$
Check: $7 \times 9 = 63$

Example 6
$$\begin{array}{r}705\\9\overline{)6345}\\-63\\04\\-0\\45\\-45\\0\end{array}$$
Check: $705 \times 9 = 6345$

Example 10
$$\begin{array}{r}3{,}058\ r3\\7\overline{)21{,}409}\\-21\\04\\-0\\40\\-35\\59\\-56\\3\end{array}$$
Check: $(3058 \times 7) + 3 =$
$21{,}406 + 3 = 21{,}409$

Example 4
$$\begin{array}{r}453\\9\overline{)4077}\\-36\\47\\-45\\27\\-27\\0\end{array}$$
Check: $453 \times 9 = 4077$

Example 8
$$\begin{array}{r}870\ r5\\6\overline{)5225}\\-48\\42\\-42\\05\\-0\\5\end{array}$$
Check: $(870 \times 6) + 5 =$
$5220 + 5 = 5225$

Example 12
$$\begin{array}{r}109\\42\overline{)4578}\\-42\\37\\-0\\378\\-378\\0\end{array}$$
Check: $109 \times 42 = 4578$

Example 14
$$\begin{array}{r}470\ \text{r}29\\39\overline{)18{,}359}\\-15\ 6\\\hline 2\ 75\\-2\ 73\\\hline 29\\-\ \ 0\\\hline 29\end{array}$$

Check: (470 × 39) + 29 =
18,330 + 29 = 18,359

Example 18
$$\begin{array}{r}421\ \text{r}33\\515\overline{)216{,}848}\\-206\ 0\\\hline 10\ 84\\-10\ 30\\\hline 548\\-515\\\hline 33\end{array}$$

Check: (421 × 515) + 33 =
216,815 + 33 = 216,848

Example 16
$$\begin{array}{r}62\ \text{r}111\\534\overline{)33{,}219}\\-32\ 04\\\hline 1\ 179\\-1\ 068\\\hline 111\end{array}$$

Check: (62 × 534) + 111 =
33,108 + 111 = 33,219

Example 20

Strategy To find the number of tires to be stored on each shelf, divide the number of tires (270) by the number of shelves (15).

Solution
$$\begin{array}{r}18\\15\overline{)270}\\-15\\\hline 120\\-120\\\hline 0\end{array}$$

Each shelf can store 18 tires.

Example 22

Strategy To find the number of cases produced in 8 hours:
- Find the number of cases produced in one hour by dividing the number of cans produced (12,600) by the number of cans to a case (24).
- Multiply the number of cases produced in one hour by 8.

Solution
$$\begin{array}{r}525\ \text{cases produced in one hour}\\24\overline{)12{,}600}\phantom{\text{ cases produced in one hour}}\\-12\ 0\phantom{00\text{ cases produced in one hour}}\\\hline 60\phantom{0\text{ cases produced in one hour}}\\-48\phantom{0\text{ cases produced in one hour}}\\\hline 120\phantom{\text{ cases produced in one hour}}\\-120\phantom{\text{ cases produced in one hour}}\\\hline 0\phantom{\text{ cases produced in one hour}}\end{array}$$

$$\begin{array}{r}525\\\times\ \ \ 8\\\hline 4200\end{array}$$

In 8 hours, 4200 cases are produced.

SECTION 1.6 *pages 47–48*

Example 2	$2^4 \cdot 3^3$	**Example 4**	10^7
Example 6	$(2 \cdot 2 \cdot 2) \cdot (5 \cdot 5) = 8 \cdot 25 = 200$	**Example 8**	$5 \cdot 4 \div 4 - 2$
			$20 \div 4 - 2$
			$5 - 2$
			3

SECTION 1.7 *pages 51–52*

Example 2	1, 2, 4, 5, 8, 10, 20 and 40 are factors of 40.	**Example 4**	$44 = 2 \cdot 2 \cdot 11$
Example 6	$177 = 3 \cdot 59$		

SOLUTIONS to Chapter 2 Examples

SECTION 2.1 *pages 63–64*

Example 2

	2	3	5	7
50 =	2		(5·5)	
84 =	(2·2)	3		(7)
135 =		(3·3·3)	5	

The LCM = $2 \cdot 2 \cdot 3 \cdot 3 \cdot 3 \cdot 5 \cdot 5 \cdot 7$
= 18,900

Example 4

	2	3	5
36 =	(2·2)	3·3	
60 =	2·2	(3)	5
72 =	2·2·2	3·3	

The GCF = $2 \cdot 2 \cdot 3 = 12$

Example 6

	2	3	5	7	11
11 =					11
24 =	2·2·2	3			
30 =	2	3	5		

Since no numbers are circled, the GCF = 1.

SECTION 2.2 *pages 67–68*

Example 2	$4\frac{1}{4}$	**Example 4**	$\frac{17}{4}$
Example 6	$4\frac{2}{5}$	**Example 8**	4
Example 10	$\frac{117}{8}$		

SECTION 2.3 *pages 71–72*

Example 2 $45 \div 5 = 9 \qquad \dfrac{3}{5} = \dfrac{3 \cdot 9}{5 \cdot 9} = \dfrac{27}{45}$

$\dfrac{27}{45}$ is equivalent to $\dfrac{3}{5}$.

Example 4 Write 6 as $\dfrac{6}{1}$.

$18 \div 1 = 18 \qquad 6 = \dfrac{6 \cdot 18}{1 \cdot 18} = \dfrac{108}{18}$

$\dfrac{108}{18}$ is equivalent to 6.

Example 6 $\dfrac{16}{24} = \dfrac{\cancel{2} \cdot \cancel{2} \cdot \cancel{2} \cdot 2}{\cancel{2} \cdot \cancel{2} \cdot \cancel{2} \cdot 3} = \dfrac{2}{3}$

Example 8 $\dfrac{8}{56} = \dfrac{\cancel{2} \cdot \cancel{2} \cdot \cancel{2}}{\cancel{2} \cdot \cancel{2} \cdot \cancel{2} \cdot 7} = \dfrac{1}{7}$

Example 10 $\dfrac{15}{32} = \dfrac{3 \cdot 5}{2 \cdot 2 \cdot 2 \cdot 2 \cdot 2} = \dfrac{15}{32}$

Example 12 $\dfrac{48}{36} = \dfrac{\cancel{2} \cdot \cancel{2} \cdot 2 \cdot 2 \cdot \cancel{3}}{\cancel{2} \cdot \cancel{2} \cdot \cancel{3} \cdot 3} = \dfrac{4}{3} = 1\dfrac{1}{3}$

SECTION 2.4 *pages 75–78*

Example 2

$\dfrac{3}{8}$
$+\dfrac{7}{8}$
$\overline{\dfrac{10}{8} = \dfrac{5}{4} = 1\dfrac{1}{4}}$

Example 4

$\dfrac{5}{12} = \dfrac{20}{48}$
$+\dfrac{9}{16} = \dfrac{27}{48}$
$\overline{\qquad \dfrac{47}{48}}$

Example 6

$\dfrac{7}{8} = \dfrac{105}{120}$
$+\dfrac{11}{15} = \dfrac{88}{120}$
$\overline{\qquad \dfrac{193}{120} = 1\dfrac{73}{120}}$

Example 8

$\dfrac{3}{4} = \dfrac{30}{40}$
$\dfrac{4}{5} = \dfrac{32}{40}$
$+\dfrac{5}{8} = \dfrac{25}{40}$
$\overline{\qquad \dfrac{87}{40} = 2\dfrac{7}{40}}$

Example 10 $7 + \dfrac{6}{11} = 7\dfrac{6}{11}$

Example 12

29
$+17\dfrac{5}{12}$
$\overline{46\dfrac{5}{12}}$

Example 14

$7\dfrac{4}{5} = 7\dfrac{24}{30}$
$6\dfrac{7}{10} = 6\dfrac{21}{30}$
$+13\dfrac{11}{15} = 13\dfrac{22}{30}$
$\overline{\qquad\qquad 26\dfrac{67}{30} = 28\dfrac{7}{30}}$

Example 16

$9\dfrac{3}{8} = 9\dfrac{45}{120}$
$17\dfrac{7}{12} = 17\dfrac{70}{120}$
$+10\dfrac{14}{15} = 10\dfrac{112}{120}$
$\overline{\qquad\qquad 36\dfrac{227}{120} = 37\dfrac{107}{120}}$

Example 18

Strategy To find the total time spent on the activities, add the three times $\left(4\frac{1}{2}, 3\frac{3}{4}, 1\frac{1}{3}\right)$.

Solution
$$4\frac{1}{2} = 4\frac{6}{12}$$
$$3\frac{3}{4} = 3\frac{9}{12}$$
$$+1\frac{1}{3} = 1\frac{4}{12}$$
$$8\frac{19}{12} = 9\frac{7}{12}$$

The total time spent on the three activities was $9\frac{7}{12}$ hours.

Example 20

Strategy To find the overtime pay:
- Find the total number of overtime hours $\left(1\frac{2}{3} + 3\frac{1}{3} + 2\right)$.
- Multiply the total number of hours by the overtime hourly wage ($24).

Solution
$$1\frac{2}{3}$$
$$3\frac{1}{3}$$
$$+2$$
$$6\frac{3}{3} = 7 \text{ hours}$$

$$\begin{array}{r} \$24 \\ \times\ 7 \\ \hline \$168 \end{array}$$

The carpenter earned $168 in overtime pay.

SECTION 2.5   pages 83–86

Example 2
$$\frac{16}{27}$$
$$-\frac{7}{27}$$
$$\frac{9}{27} = \frac{1}{3}$$

Example 6
$$\frac{13}{18} = \frac{52}{72}$$
$$-\frac{7}{24} = \frac{21}{72}$$
$$\frac{31}{72}$$

Example 10
$$8 \phantom{\frac{4}{13}} = 7\frac{13}{13}$$
$$-2\frac{4}{13} = 2\frac{4}{13}$$
$$5\frac{9}{13}$$

Example 14
$$81\frac{17}{80} = 80\frac{97}{80}$$
$$-17\frac{29}{80} = 17\frac{29}{80}$$
$$63\frac{68}{80} = 63\frac{17}{20}$$

Example 4
$$\frac{5}{6} = \frac{20}{24}$$
$$-\frac{3}{8} = \frac{9}{24}$$
$$\frac{11}{24}$$

Example 8
$$17\frac{5}{9} = 17\frac{20}{36}$$
$$-11\frac{5}{12} = 11\frac{15}{36}$$
$$6\frac{5}{36}$$

Example 12
$$21\frac{7}{9} = 21\frac{28}{36} = 20\frac{64}{36}$$
$$-\ 7\frac{11}{12} = 7\frac{33}{36} = 7\frac{33}{36}$$
$$13\frac{31}{36}$$

Example 16

Strategy To find the time remaining before the plane lands, subtract the number of hours already in the air $\left(2\frac{3}{4}\right)$ from the total time of the trip $\left(5\frac{1}{2}\right)$.

Example 18

Strategy To find the amount of weight to be lost during the third month:
- Find the total weight loss during the first two months $\left(7\frac{1}{2} + 5\frac{3}{4}\right)$.
- Subtract the total weight loss from the goal (24 pounds).

Solution
$$5\tfrac{1}{2} = 5\tfrac{2}{4} = 4\tfrac{6}{4}$$
$$-2\tfrac{3}{4} = 2\tfrac{3}{4} = 2\tfrac{3}{4}$$
$$\overline{2\tfrac{3}{4}\text{ hours}}$$

The plane will land in $2\tfrac{3}{4}$ hours.

Solution
$$7\tfrac{1}{2} = 7\tfrac{2}{4}$$
$$+5\tfrac{3}{4} = 5\tfrac{3}{4}$$
$$\overline{12\tfrac{5}{4} = 13\tfrac{1}{4}\text{ pounds lost}}$$

$$24 = 23\tfrac{4}{4}$$
$$-13\tfrac{1}{4} = 13\tfrac{1}{4}$$
$$\overline{10\tfrac{3}{4}\text{ pounds}}$$

The patient must lose $10\tfrac{3}{4}$ pounds to achieve the goal.

SECTION 2.6   *pages 91–94*

Example 2
$$\frac{12}{25} \times \frac{5}{16} = \frac{12 \cdot 5}{25 \cdot 16} = \frac{\cancel{2}\cdot\cancel{2}\cdot 3\cdot\cancel{5}}{5\cdot\cancel{5}\cdot\cancel{2}\cdot\cancel{2}\cdot 2\cdot 2} = \frac{3}{20}$$

Example 4
$$\frac{4}{21} \times \frac{7}{44} = \frac{4 \cdot 7}{21 \cdot 44} = \frac{\cancel{2}\cdot\cancel{2}\cdot\cancel{7}}{3\cdot\cancel{7}\cdot\cancel{2}\cdot\cancel{2}\cdot 11} = \frac{1}{33}$$

Example 6
$$\frac{2}{21} \times \frac{10}{33} = \frac{2 \cdot 10}{21 \cdot 33} = \frac{2\cdot 2\cdot 5}{3\cdot 7\cdot 3\cdot 11} = \frac{20}{693}$$

Example 8
$$\frac{16}{5} \times \frac{15}{24} = \frac{16 \cdot 15}{5 \cdot 24} = \frac{\cancel{2}\cdot\cancel{2}\cdot\cancel{2}\cdot 2\cdot\cancel{3}\cdot\cancel{5}}{\cancel{5}\cdot\cancel{2}\cdot\cancel{2}\cdot\cancel{2}\cdot\cancel{3}} = 2$$

Example 10
$$5\tfrac{2}{5} \times \frac{5}{9} = \frac{27}{5} \times \frac{5}{9} = \frac{27 \cdot 5}{5 \cdot 9} = \frac{\cancel{3}\cdot\cancel{3}\cdot 3\cdot\cancel{5}}{\cancel{5}\cdot\cancel{3}\cdot\cancel{3}} = 3$$

Example 12
$$9 \times \frac{7}{12} = \frac{9}{1} \times \frac{7}{12} = \frac{9 \cdot 7}{1 \cdot 12} = \frac{\cancel{3}\cdot 3\cdot 7}{1\cdot 2\cdot 2\cdot\cancel{3}} = \frac{21}{4} = 5\tfrac{1}{4}$$

Example 14
$$3\tfrac{2}{5} \times 6\tfrac{1}{4} = \frac{17}{5} \times \frac{25}{4} = \frac{17 \cdot 25}{5 \cdot 4} = \frac{17\cdot\cancel{5}\cdot 5}{\cancel{5}\cdot 2\cdot 2} = \frac{85}{4} = 21\tfrac{1}{4}$$

Example 16
$$3\tfrac{2}{7} \times 6 = \frac{23}{7} \times \frac{6}{1} = \frac{23 \cdot 6}{7 \cdot 1} = \frac{23 \cdot 3 \cdot 2}{7 \cdot 1} = \frac{138}{7} = 19\tfrac{5}{7}$$

Example 18

Strategy   To find the value of the house today, multiply the old value of the house ($30,000) by $3\tfrac{1}{2}$.

Solution $\dfrac{30{,}000}{1} \times \dfrac{7}{2} = \dfrac{30{,}000 \cdot 7}{1 \cdot 2} = 105{,}000$

The value of the house today is $105,000.

Example 20

Strategy   To find the cost of the air compressor:
- Multiply to find the value of the drying chamber $\left(\tfrac{4}{5} \times \$60{,}000\right)$.
- Subtract the value of the drying chamber from the total value of the two items ($60,000).

Solution $\dfrac{4}{5} \times \dfrac{\$60{,}000}{1} = \dfrac{\$240{,}000}{5} = \$48{,}000$

value of the drying chamber

$60,000
−48,000
$12,000

The cost of the air compressor is $12,000.

SECTION 2.7 pages 99–102

Example 2 $\frac{3}{7} \div \frac{2}{3} = \frac{3}{7} \times \frac{3}{2} = \frac{3 \cdot 3}{7 \cdot 2} = \frac{9}{14}$

Example 4 $\frac{3}{4} \div \frac{9}{10} = \frac{3}{4} \times \frac{10}{9} = \frac{3 \cdot 10}{4 \cdot 9} =$
$\frac{\cancel{3} \cdot \cancel{2} \cdot 5}{\cancel{2} \cdot 2 \cdot \cancel{3} \cdot 3} = \frac{5}{6}$

Example 6 $8 \div \frac{2}{3} = \frac{8}{1} \div \frac{2}{3} = \frac{8}{1} \times \frac{3}{2} =$
$\frac{\cancel{2} \cdot 2 \cdot 2 \cdot 3}{\cancel{2}} = 12$

Example 8 $\frac{5}{7} \div 6 = \frac{5}{7} \div \frac{6}{1} =$
$\frac{5}{7} \times \frac{1}{6} = \frac{5}{7 \cdot 2 \cdot 3} = \frac{5}{42}$

Example 10 $12\frac{3}{5} \div 7 = \frac{63}{5} \div \frac{7}{1} = \frac{63}{5} \times \frac{1}{7} =$
$\frac{63 \cdot 1}{5 \cdot 7} = \frac{\cdot 3 \cdot 3 \cdot \cancel{7}}{5 \cdot \cancel{7}} = \frac{9}{5} = 1\frac{4}{5}$

Example 12 $3\frac{2}{3} \div 2\frac{2}{5} = \frac{11}{3} \div \frac{12}{5} =$
$\frac{11}{3} \times \frac{5}{12} = \frac{11 \cdot 5}{3 \cdot 12} =$
$\frac{11 \cdot 5}{3 \cdot 2 \cdot 2 \cdot 3} = \frac{55}{36} = 1\frac{19}{36}$

Example 14 $2\frac{5}{6} \div 8\frac{1}{2} = \frac{17}{6} \div \frac{17}{2} = \frac{17}{6} \times \frac{2}{17} =$
$\frac{17 \cdot 2}{6 \cdot 17} = \frac{\cancel{17} \cdot \cancel{2}}{\cancel{2} \cdot 3 \cdot \cancel{17}} = \frac{1}{3}$

Example 16 $6\frac{2}{5} \div 4 = \frac{32}{5} \div \frac{4}{1} = \frac{32}{5} \times \frac{1}{4} =$
$\frac{32 \cdot 1}{5 \cdot 4} = \frac{2 \cdot 2 \cdot 2 \cdot \cancel{2} \cdot \cancel{2}}{5 \cdot \cancel{2} \cdot \cancel{2}} = \frac{8}{5} = 1\frac{3}{5}$

Example 18

Strategy To find the price of one ounce of gold, divide the total price of the coin ($975) by the number of ounces $\left(2\frac{1}{2}\right)$.

Solution $\$975 \div 2\frac{1}{2} = \frac{\$975}{1} \div \frac{5}{2} =$
$\frac{\$975}{1} \times \frac{2}{5} = \frac{\$975 \cdot 2}{1 \cdot 5} = \$390$

The price of one ounce of gold is $390.

Example 20

Strategy To find the length of the remaining piece:
- Divide the total length of the board (16 feet) by the length of each shelf $\left(3\frac{1}{3}\text{ feet}\right)$.
- Multiply the fractional part of the result by the length of one shelf to determine the length of the remaining piece.

Solution $16 \div 3\frac{1}{3} = 16 \div \frac{10}{3} =$
$\frac{16}{1} \times \frac{3}{10} = \frac{16 \cdot 3}{1 \cdot 10} = \frac{\cancel{2} \cdot 2 \cdot 2 \cdot 2 \cdot 3}{\cancel{2} \cdot 5} =$
$\frac{24}{5} = 4\frac{4}{5}$

$\frac{4}{5} \times \frac{10}{3} = \frac{40}{15} = 2\frac{2}{3}$

The length of the piece remaining is $2\frac{2}{3}$ feet.

SECTION 2.8 pages 107–108

Example 2 $\dfrac{9}{14} > \dfrac{13}{21}$

Example 6 $\left(\dfrac{1}{13}\right)^2 \cdot \left(\dfrac{1}{4} + \dfrac{1}{6}\right) \div \dfrac{5}{13}$

$\left(\dfrac{1}{13}\right)^2 \cdot \left(\dfrac{5}{12}\right) \div \dfrac{5}{13}$

$\left(\dfrac{1}{169}\right) \cdot \left(\dfrac{5}{12}\right) \div \dfrac{5}{13}$

$\left(\dfrac{1 \cdot 5}{13 \cdot 13 \cdot 12}\right) \div \dfrac{5}{13}$

$\dfrac{1 \cdot \cancel{5} \cdot \cancel{13}}{\cancel{13} \cdot 13 \cdot 12 \cdot \cancel{5}}$

$\dfrac{1}{156}$

Example 4 $\left(\dfrac{7}{11}\right)^2 \cdot \left(\dfrac{2}{7}\right) = \left(\dfrac{7}{11} \cdot \dfrac{7}{11}\right) \cdot \left(\dfrac{2}{7}\right) =$
$\dfrac{7 \cdot 7 \cdot 2}{11 \cdot 11 \cdot 7} = \dfrac{14}{121}$

SOLUTIONS to Chapter 3 Examples

SECTION 3.1 pages 121–122

Example 2 Two hundred nine and five thousand eight hundred thirty-eight hundred-thousandths

Example 6 4.35

Example 4 42,000.000207

Example 8 3.29053

SECTION 3.2 pages 125–126

Example 2
```
  1 2
   4.62
  27.9
+  0.62054
  33.14054
```

Example 6

Strategy To find the total cost of the four items, add the four costs ($850.65, $282.97, $650.00, and $835.00).

Solution
```
  $850.65
   282.97
   650.00
 + 835.00
  $2618.62
```

The total cost of the four items is $2618.62.

Example 4
```
     1
   6.05
  12.
+  0.374
  18.424
```

Example 8

Strategy To find the total income, add the four commissions ($485.60, $599.46, $326.75, and $725.42) to the salary ($425.00).

Solution
```
  $485.60
   599.46
   326.75
   725.42
 + 425.00
  $2562.23
```

The executive's total income was $2562.23.

SECTION 3.3 *pages 129–130*

Example 2

$$\begin{array}{r} \overset{119}{6\cancel{1}\cancel{10}13}\\ 7\,2.\,0\,3\,9 \\ -8.\,4\,7 \\ \hline 6\,3.\,5\,6\,9 \end{array}$$

Check:
$$\begin{array}{r} {\scriptstyle 1\,1\,1}\\ 8.47 \\ +63.569 \\ \hline 72.039 \end{array}$$

Example 4

$$\begin{array}{r} \overset{149}{2\cancel{4}\cancel{10}10}\\ 3\,5.\,0\,0 \\ -9.\,6\,7 \\ \hline 2\,5.\,3\,3 \end{array}$$

Check:
$$\begin{array}{r} {\scriptstyle 1\,1\,1}\\ 9.67 \\ +25.33 \\ \hline 35.00 \end{array}$$

Example 6

$$\begin{array}{r} \overset{16\,9\,9}{2\cancel{6}\cancel{10}\cancel{10}10}\\ 3.\,7\,0\,0\,0 \\ -1.\,9\,7\,1\,5 \\ \hline 1.\,7\,2\,8\,5 \end{array}$$

Check:
$$\begin{array}{r} {\scriptstyle 1\,\,111}\\ 1.9715 \\ +1.7285 \\ \hline 3.7000 \end{array}$$

Example 8

Strategy To find the amount of change, subtract the amount paid ($3.85) from $5.00.

Solution
$$\begin{array}{r} \$5.00 \\ -3.85 \\ \hline \$1.15 \end{array}$$

Your change is $1.15.

Example 10

Strategy To find the new balance.
- Add to find the total of the three checks ($1025.60 + $79.85 + $162.47).
- Subtract the total from the previous balance ($2472.69).

Solution
$$\begin{array}{r} \$1025.60 \\ 79.85 \\ +162.47 \\ \hline \$1267.92 \end{array} \qquad \begin{array}{r} \$2472.69 \\ -1267.92 \\ \hline \$1204.77 \end{array}$$

The new balance is $1204.77.

SECTION 3.4 *pages 133–136*

Example 2
$$\begin{array}{r} 870 \\ \times\;\;4.6 \\ \hline 5220 \\ 3480 \\ \hline 4002.0 \end{array}$$

Example 4
$$\begin{array}{r} 0.28 \\ \times\;\;0.7 \\ \hline 0.196 \end{array}$$

Example 6
$$\begin{array}{r} 0.000086 \\ \times\;\;\;\;0.057 \\ \hline 602 \\ 430 \\ \hline 0.000004902 \end{array}$$

Example 8
$$\begin{array}{r} 79.063 \\ \times\;\;\;\;0.47 \\ \hline 553441 \\ 316252 \\ \hline 37.15961 \end{array}$$

Example 10
$$\begin{array}{r} 4.68 \\ \times 6.03 \\ \hline 1404 \\ 28\,080 \\ \hline 28.2204 \end{array}$$

Example 12 $6.9 \times 1000 = 6900$

Example 14 $4.0273 \times 10^2 = 402.73$

Example 16

Strategy To find the cost of running the freezer for 210 hours, multiply the hourly cost ($.035) by the number of hours (210).

Solution
$.035
×210
─────
$7.35

The cost of running the freezer for 210 hours is $7.35.

Example 18

Strategy To find the total cost of the stereo:
- Multiply the monthly payment ($37.18) by the number of months (18).
- Add the total to the down payment ($175.00).

Solution
$37.18 $699.24
×   18   + 175.00
──────   ────────
$669.24 $844.24

SECTION 3.5 pages 141–144

Example 2

$$\begin{array}{r}2.7\\0.052\overline{)0.140.4}\\-104\\\hline 36\ 4\\-36\ 4\\\hline 0\end{array}$$

Example 4

$$\begin{array}{r}0.4873 \approx 0.487\\76\overline{)37.0420}\\-30\ 4\\\hline 6\ 64\\-6\ 08\\\hline 562\\-532\\\hline 300\\-228\end{array}$$

Example 6

$$\begin{array}{r}72.73 \approx 72.7\\5.09\overline{)370.20.00}\\-356\ 3\\\hline 13\ 90\\-10\ 18\\\hline 3\ 720\\-3\ 563\\\hline 1570\\-1527\end{array}$$

Example 8 $309.21 \div 10{,}000 = 0.030921$

Example 10 $42.93 \div 10^4 = 0.004293$

Example 12

Strategy To find the trucker's monthly income, divide the annual income ($23,741.40) by the number of months (12).

Example 14

Strategy To find the store's total income:
- Divide the total amount paid in taxes ($138.60) by the tax on each lamp ($11.55).
- Multiply the quotient by the price of one lamp ($95.60).

Solution	$\quad\quad\quad\quad\;\; \$1,978.45$ $12\overline{)\;\$23,741.40}$ $\quad\underline{-12}$ $\quad\quad 11\;7$ $\quad\underline{-10\;8}$ $\quad\quad\quad 94$ $\quad\quad\underline{-84}$ $\quad\quad\quad 101$ $\quad\quad\underline{-\;\;96}$ $\quad\quad\quad\quad 5\;4$ $\quad\quad\quad\underline{-4\;8}$ $\quad\quad\quad\quad\;\; 60$ $\quad\quad\quad\;\;\underline{-60}$ $\quad\quad\quad\quad\quad\; 0$ The trucker's monthly income is $1978.45.	**Solution**	$\quad\quad\quad\quad\quad 12.$ $\$11.55\overline{)\$138.60.}$ $\quad\quad\underline{-115\;5}$ $\quad\quad\quad\;\; 23\;10$ $\quad\quad\;\;\underline{-23\;10}$ $\quad\quad\quad\quad\quad\; 0$ $\quad\quad\;\;\$95.60$ $\quad\underline{\times\quad\quad 12}$ $\quad\$1147.20$ The store's total income was $1147.20.

SECTION 3.6 pages 149–150

Example 2	$\quad 0.56 \approx 0.6$ $16\overline{)9.00}$	**Example 4**	$4\frac{1}{6} = \frac{25}{6}\quad\quad 4.166 \approx 4.17$ $\quad\quad\quad\quad\; 6\overline{)25.00}$
Example 6	$0.56 = \frac{56}{100} = \frac{14}{25}$ $5.35 = 5\frac{35}{100} = 5\frac{7}{20}$	**Example 8**	$0.12\frac{7}{8} = \frac{12\frac{7}{8}}{100} = 12\frac{7}{8} \div 100 =$ $\frac{103}{8} \times \frac{1}{100} = \frac{103}{800}$
Example 10	$64.009 < 64.99$	**Example 12**	$0.63 > \frac{5}{8}$

SOLUTIONS to Chapter 4 Examples

SECTION 4.1 pages 163–164

Example 2	$\frac{20 \text{ pounds}}{24 \text{ pounds}} = \frac{20}{24} = \frac{5}{6}$ 20 pounds : 24 pounds = 20 : 24 = 5 : 6 20 pounds TO 24 pounds = 20 TO 24 = 5 TO 6	**Example 4**	$\frac{64 \text{ miles}}{8 \text{ miles}} = \frac{64}{8} = \frac{8}{1}$ 64 miles : 8 miles = 64 : 8 = 8 : 1 64 miles TO 8 miles = 64 TO 8 = 8 TO 1
Example 6		**Example 8**	
Strategy	To find the ratio, write the ratio of board feet of cedar (12,000) to board feet of ash (18,000) in simplest form.	**Strategy**	To find the ratio, write the ratio of the amount spent on radio advertising ($15,000) to the amount spent on radio and television advertising ($15,000 + $20,000).
Solution	$\frac{12,000}{18,000} = \frac{2}{3}$ The ratio is $\frac{2}{3}$.	**Solution**	$\frac{\$15,000}{\$15,000 + \$20,000} = \frac{\$15,000}{\$35,000} = \frac{3}{7}$ The ratio is $\frac{3}{7}$.

SECTION 4.2 *pages 167–168*

Example 2 $\dfrac{15 \text{ pounds}}{12 \text{ trees}} = \dfrac{5 \text{ pounds}}{4 \text{ trees}}$

Example 4 $\dfrac{260 \text{ miles}}{8 \text{ hours}} = \dfrac{32.5 \text{ miles}}{1 \text{ hour}}$
  $= 32.5 \text{ miles/hour}$

Example 6

Strategy To find the miles per hour rate, divide the number of miles (47) by the number of hours (3).

Solution
$$3\overline{)47.00} = 15.66$$

The rate is 15.7 miles/hour.

Example 8

Strategy To find the jeweler's profit per ounce:
- Find the total profit by subtracting the cost ($1980) from the selling price ($2075).
- Find the profit per ounce by dividing the total profit by the number of ounces (5).

Solution
$2075
-1980
$\overline{\$\ 110}$ total profit

$5\overline{)\$95} = \$19$

The profit per ounce is $19.

SECTION 4.3 *pages 171–174*

Example 2 $\dfrac{6}{10} \bowtie \dfrac{9}{15}$ $\rightarrow 10 \times 9 = 90$
  $\rightarrow 6 \times 15 = 90$
The proportion is true.

Example 4 $\dfrac{32}{6} \bowtie \dfrac{90}{8}$ $\rightarrow 6 \times 90 = 540$
  $\rightarrow 32 \times 8 = 256$
The proportion is not true.

Example 6
$n \times 7 = 14 \times 3$
$n \times 7 = 42$
$n = 42 \div 7$
$n = 6$

Check: $\dfrac{6}{14} \bowtie \dfrac{3}{7}$ $\rightarrow 14 \times 3 = 42$
  $\rightarrow 6 \times 7 = 42$

Example 8
$5 \times 20 = 8 \times n$
$100 = 8 \times n$
$100 \div 8 = n$
$12.5 = n$

Example 10
$15 \times n = 20 \times 12$
$15 \times n = 240$
$n = 240 \div 15$
$n = 16$

Check: $\dfrac{15}{20} \bowtie \dfrac{12}{16}$ $\rightarrow 20 \times 12 = 240$
  $\rightarrow 15 \times 16 = 240$

Example 12
$12 \times 4 = 7 \times n$
$48 = 7 \times n$
$48 \div 7 = n$
$6.86 \approx n$

Example 14
$n \times 1 = 12 \times 4$
$n \times 1 = 48$
$n = 48 \div 1$
$n = 48$

Check: $\dfrac{48}{12} \bowtie \dfrac{4}{1}$ $\rightarrow 12 \times 4 = 48$
  $\rightarrow 48 \times 1 = 48$

Example 16
$3 \times n = 12 \times 8$
$3 \times n = 96$
$n = 96 \div 3$
$n = 32$

Check: $\dfrac{3}{8} \bowtie \dfrac{12}{32}$ $\rightarrow 8 \times 12 = 96$
  $\rightarrow 3 \times 32 = 96$

Example 18

Strategy To find the number of jars that can be packed in 15 boxes, write and solve a proportion using n to represent the number of jars.

Example 20

Strategy To find the number of tablespoons of fertilizer needed, write and solve a proportion using n to represent the number of tablespoons of fertilizer.

Solution	$\dfrac{24 \text{ jars}}{6 \text{ boxes}} = \dfrac{n \text{ jars}}{15 \text{ boxes}}$	**Solution**	$\dfrac{3 \text{ tablespoons}}{4 \text{ gallons}} = \dfrac{n \text{ tablespoons}}{10 \text{ gallons}}$

$$24 \times 15 = 6 \times n$$
$$360 = 6 \times n$$
$$360 \div 6 = n$$
$$60 = n$$

60 jars can be packed in 15 boxes.

$$3 \times 10 = 4 \times n$$
$$30 = 4 \times n$$
$$30 \div 4 = n$$
$$7\tfrac{1}{2} = n$$

For 10 gallons of water, $7\tfrac{1}{2}$ tablespoons of fertilizer are required.

SOLUTIONS to Chapter 5 Examples

SECTION 5.1 pages 189–190

Example 2 $125\% = 125 \times \dfrac{1}{100} = \dfrac{125}{100} = 1\tfrac{1}{4}$

$125\% = 125 \times 0.01 = 1.25$

Example 4 $33\tfrac{1}{3}\% = 33\tfrac{1}{3} \times \dfrac{1}{100} = \dfrac{100}{3} \times \dfrac{1}{100}$

$= \dfrac{100}{300} = \dfrac{1}{3}$

Example 6 $0.25\% = 0.25 \times 0.01 = 0.0025$

Example 8 $0.048 = 0.048 \times 100\% = 4.8\%$

Example 10 $3.67 = 3.67 \times 100\% = 367\%$

Example 12 $0.62\tfrac{1}{2} = 0.62\tfrac{1}{2} \times 100\% = 62\tfrac{1}{2}\%$

Example 14 $\dfrac{5}{6} = \dfrac{5}{6} \times 100\% = \dfrac{500\%}{6} = 83\tfrac{1}{3}\%$

Example 16 $1\tfrac{4}{9} = \dfrac{13}{9} = \dfrac{13}{9} \times 100\%$

$= \dfrac{1300\%}{9} \approx 144.4\%$

SECTION 5.2 pages 193–194

Example 2 $n = 0.063 \times 150$
$n = 9.45$

Example 4 $n = \dfrac{1}{6} \times 66$
$n = 11$

Example 6

Strategy To find the sales tax, write and solve a basic percent equation using n to represent the sales tax (amount). The percent is 6% and the base is $450.

Solution $6\% \times \$450 = n$
$.06 \times \$450 = n$
$\$27 = n$

The sales tax was $27.

Example 8

Strategy To find the new hourly wage:
- Find the amount of the raise. Write and solve a basic percent equation using n to represent the amount of the raise (amount). The percent is 8%. The base is $13.50.
- Add the amount of the raise to the old wage.

Solution $8\% \times \$13.50 = n$ $\$13.50$
$0.08 \times \$13.50 = n$ $+\ 1.08$
$\$1.08 = n$ $\overline{\$14.58}$

The new hourly wage is $14.58.

SECTION 5.3 pages 197–198

Example 2
$$n \times 32 = 16$$
$$n = 16 \div 32$$
$$n = 0.50$$
$$n = 50\%$$

Example 6
$$30 = n \times 45$$
$$30 \div 45 = n$$
$$0.66\tfrac{2}{3} = n$$
$$66\tfrac{2}{3}\% = n$$

Example 4
$$n \times 15 = 48$$
$$n = 48 \div 15$$
$$n = 3.20$$
$$n = 320\%$$

Example 8

Strategy To find what percent of the income the income tax is, write and solve a basic percent equation using n to represent the percent. The base is $20,000 and the amount is $3000.

Solution
$$n \times \$20{,}000 = \$3000$$
$$n = \$3000 \div \$20{,}000$$
$$n = 0.15 = 15\%$$

The income tax is 15% of the income.

Example 10

Strategy To find the percent that did not favor the candidate:
- Subtract to find the number of people that did not favor the candidate (1000 − 667).
- Write and solve a basic percent equation using n to represent the percent. The base is 1000 and the amount is the number of people who did not favor the candidate.

Solution 1000 − 667 = 333   number who did not favor the candidate
$$n \times 1000 = 333$$
$$n = 333 \div 1000$$
$$n = 0.333 = 33.3\%$$

33.3% of the people did not favor the candidate.

SECTION 5.4 pages 201–202

Example 2
$$0.86 \times n = 215$$
$$n = 215 \div 0.86$$
$$n = 250$$

Example 6
$$\tfrac{1}{6} \times n = 5$$
$$n = 5 \div \tfrac{1}{6}$$
$$n = 30$$

Example 4
$$0.025 \times n = 15$$
$$n = 15 \div 0.025$$
$$n = 600$$

Example 8

Strategy To find the original value of the car, write and solve a basic percent equation using n to represent the original value (base). The percent is 51%. The amount is $3876.

Example 10

Strategy To find the difference between the original price and the selling price:
- Find the original price. Write and solve a basic percent equation using n to represent the original price (base). The percent is 80%. The amount is $44.80.
- Subtract the sale price ($44.80) from the original price.

Solution	$51\% \times n = \$3876$	**Solution**	$80\% \times n = \$44.80$
	$0.51 \times n = \$3876$		$0.80 \times n = \$44.80$
	$n = \$3876 \div 51$		$n = \$44.80 \div 0.80$
	$n = \$7600$		$n = \$56.00$ original price

The original value of the car was $7600.

$56.00 - $44.80 = $11.20

The difference between the original price and the sale price is $11.20.

SECTION 5.5 pages 205–206

Example 2

$$\frac{26}{100} = \frac{22}{n}$$

$26 \times n = 100 \times 22$
$26 \times n = 2200$
$n = 2200 \div 26$
$n \approx 84.62$

Example 4

$$\frac{16}{100} = \frac{n}{132}$$

$16 \times 132 = 100 \times n$
$2112 = 100 \times n$
$2112 \div 100 = n$
$21.12 = n$

Example 6

Strategy To find the number of days it snowed, write and solve a proportion using n to represent the number of days (amount). The percent is 64%. The base is 150.

Solution

$$\frac{64}{100} = \frac{n}{150}$$

$64 \times 150 = 100 \times n$
$9600 = 100 \times n$
$9600 \div 100 = n$
$96 = n$

It snowed 96 days.

Example 8

Strategy To find the percent of pens which were not defective:
- Subtract to find the number of pens which were not defective (200 − 5).
- Write and solve a proportion using n to represent the percent of pens which were not defective. The base is 200 and the amount is the number of pens not defective.

Solution $200 - 5 = 195$ number of pens not defective

$$\frac{n}{100} = \frac{195}{200}$$

$200 \times n = 195 \times 100$
$200 \times n = 19500$
$n = 19500 \div 200$
$n = 97.5\%$

97.5% of the pens were not defective.

SOLUTIONS to Chapter 6 Examples

SECTION 6.1   pages 219–220

Example 2   a. $6.25 \div 5 = 1.25$
      Unit cost: $1.25 per quart
   b. $85 \div 4 = 21.25$
      Unit cost: 21.3¢ per ear

Example 4

Strategy   To find the more economical purchase, compare unit costs.

Solution $2.52 \div 6 = 0.42$
   $1.66 \div 4 = 0.415$
   $\$.415 < \$.42$

   The more economical purchase is 4 cans for $1.66.

Example 6

Strategy   To find the total cost, multiply the unit cost ($4.96) by the number of units (7).

Solution $4.96 \times 7 = 34.72$

   The total cost is $34.72.

SECTION 6.2   pages 223–228

Example 2

Strategy   To find the percent increase:
- Find the amount of increase.
- Solve the basic percent equation for *percent*.

Solution
$$\begin{array}{r} 2250 \\ -2000 \\ \hline 250 \end{array}$$
   $n \times 2000 = 250$
   $n = 250 \div 2000$
   $n = 0.125 = 12.5\%$

   The percent increase was 12.5%.

Example 4

Strategy   To find the new hourly wage:
- Solve the basic percent equation for *amount*.
- Add the amount of increase to the original wage.

Solution $0.14 \times 6.50 = n$
   $0.91 = n$
   $6.50 + 0.91 = 7.41$

   The new hourly wage is $7.41.

Example 6

Strategy   To find the markup, solve the basic percent equation for *amount*.

Solution $0.20 \times 8 = n$
   $1.60 = n$

   The markup is $1.60.

Example 8

Strategy   To find the selling price:
- Find the markup by solving the basic percent equation for *amount*.
- Add the markup to the cost.

Solution $0.55 \times 72 = n$
   $39.60 = n$
   $72 + 39.60 = 111.60$

   The selling price is $111.60.

Example 10

Strategy   To find the percent decrease:
- Find the amount of decrease.
- Solve the basic percent equation for *percent*.

Example 12

Strategy   To find the visibility:
- Find the amount of decrease by solving the basic percent equation for *amount*.
- Subtract the amount of decrease from the original visibility.

Solution	$150 - 120 = 30$ $n \times 150 = 30$ $n = 30 \div 150$ $n = 0.2 = 20\%$ The percent decrease was 20%.	**Solution**	$0.40 \times 5 = n$ $2 = n$ $5 - 2 = 3$ The visibility was 3 miles.

Example 14

Strategy To find the discount rate, solve the basic percent equation for *percent*.

Solution $n \times 125 = 25$
$n = 0.20 = 20\%$

The discount rate is 20%.

Example 16

Strategy To find the sale price:
- Find the discount by solving the basic percent equation for *amount*.
- Subtract to find the sale price.

Solution $0.20 \times 10.25 = n$
$2.05 = n$
$10.25 - 2.05 = 8.20$

The sale price is $8.20.

SECTION 6.3 *pages 233–234*

Example 2

Strategy To find the simple interest, multiply the principal by the annual interest rate by the time (in years).

Solution $15{,}000 \times 0.12 \times 1.5 = 2700$

The interest due is $2700.

Example 6

Strategy To find the interest earned:
- Find the new principal by multiplying the original principal by the factor found in the compound interest table.
- Subtract the original principal from the new principal.

Solution $1000 \times 7.20957 = 7209.57$
$7209.57 - 1000 = 6209.57$

The interest earned is $6209.57.

Example 4

Strategy To find the interest due, multiply the principal by the monthly interest rate by the time (in months).

Solution $400 \times 0.02 \times 2 = 16$

The interest charge is $16.

SECTION 6.4 pages 237–240

Example 2

Strategy To find the mortgage:
- Find the down payment by solving the basic percent equation for *amount*.
- Subtract the down payment from the purchase price.

Solution
$0.25 \times 216{,}000 = n$
$54{,}000 = n$

$216{,}000 - 54{,}000 = 162{,}000$

The mortgage is $162,000.

Example 6

Strategy To find the monthly mortgage payment:
- Subtract the down payment from the purchase price to find the mortgage.
- Multiply the mortgage by the factor found in the monthly payment table.

Solution
$75{,}000 - 15{,}000 = 60{,}000$
$60{,}000 \times 0.0110109 = 660.654$

The monthly mortgage payment is $660.65.

Example 10

Strategy To find the monthly payment:
- Divide the annual property tax by 12 to find the monthly property tax.
- Add the monthly property tax to the monthly mortgage payment.

Solution
$744 \div 12 = 62$
$415.20 + 62 = 477.20$

The total monthly payment is $477.20.

Example 4

Strategy To find the loan origination fee, solve the basic percent equation for *amount*.

Solution
$0.045 \times 80{,}000 = n$
$3600 = n$

The loan origination fee is $3600.

Example 8

Strategy To find the interest:
- Multiply the mortgage by the factor found in the monthly payment table on page A6 to find the monthly mortgage payment.
- Subtract the principal from the monthly mortgage payment.

Solution
$125{,}000 \times 0.0083920 = 1049$
$1049 - 492.65 = 556.35$

The interest on the mortgage is $556.35.

SECTION 6.5 pages 243–244

Example 2

Strategy To find the amount financed:
- Find the down payment by solving the basic percent equation for *amount*.
- Subtract the down payment from the purchase price.

Example 4

Strategy To find the license fee, solve the basic percent equation for *amount*.

Solution $0.20 \times 9200 = n$
$1840 = n$
$9200 - 1840 = 7360$

The amount financed is $7360.

Example 6

Strategy   To find the cost of operating the car, multiply the cost per mile by the number of miles driven.

Solution $23{,}000 \times 0.22 = 5060$

The cost of operating the car is $5060.

Example 10

Strategy   To find the monthly payment:
- Subtract the down payment from the purchase price to find the amount financed.
- Multiply the amount financed by the factor found in the monthly payment table on page A6.

Solution $15{,}900 - 3975 = 11{,}925$
$11{,}925 \times 0.0258455 = 308.2075$

The monthly payment is $308.21.

Solution $0.015 \times 7350 = n$
$110.25 = n$

The license fee is $110.25.

Example 8

Strategy   To find the cost per mile for the car insurance, divide the cost for insurance by the number of miles driven.

Solution $360 \div 15{,}000 = 0.024$

The cost per mile for insurance is $0.024.

SECTION 6.6   pages 247–248

Example 2

Strategy   To find the worker's earnings:
- Find the worker's overtime wage by multiplying the hourly wage by 2.
- Multiply the number of hours worked by the overtime hours.

Solution $8.50 \times 2 = 17$
$17 \times 8 = 136$

The construction worker earns $136.

Example 6

Strategy   To find the total earnings:
- Find the commission by multiplying the commission rate by the sales over $50,000.
- Add the commission to the annual salary.

Solution $175{,}000 - 50{,}000 = 125{,}000$
$125{,}000 \times 0.095 = 11{,}875$
$12{,}000 + 11{,}875 = 23{,}875$

The insurance agent earned $23,875.

Example 4

Strategy   To find the salary per month, divide the annual salary by the number of months in a year (12).

Solution $28{,}224 \div 12 = 2352$

The contractor's monthly salary is $2352.

SECTION 6.7 pages 251–256

Example 2

Strategy To find the current balance:
- Subtract the amount of the check from the old balance.
- Add the amount of each deposit.

Solution
```
   302.46
 −  20.59   check
   281.87
   176.86   first deposit
 +  94.73   second deposit
   553.46
```

The current checking account balance is $553.46.

Example 4

Solution

Current checkbook balance:	653.41
Check: 237	+ 48.73
	702.14
Interest:	+ 2.11
	704.25
Deposit:	−523.84
	180.41

Closing bank balance from bank statement: $180.41.

Checkbook balance: $180.41.

The bank statement and checkbook balance.

ANSWERS to Chapter 1 Odd-Numbered Exercises

SECTION 1.1 pages 7–8

1. (number line with point at 3, range 0–12) **3.** (number line with point at 9, range 0–12) **5.** $37 < 49$ **7.** $101 > 87$
9. $245 > 158$ **11.** $0 < 45$ **13.** $815 < 928$ **15.** $7003 < 7020$ **17.** Eight hundred five **19.** Four hundred eighty-five **21.** Two thousand six hundred seventy-five **23.** Forty-two thousand nine hundred twenty-eight **25.** Three hundred fifty-six thousand nine hundred forty-three **27.** Three million six hundred ninety-seven thousand four hundred eighty-three **29.** 85 **31.** 3456 **33.** 609,948
35. 4,003,002 **37.** $5000 + 200 + 80 + 7$ **39.** $50,000 + 8000 + 900 + 40 + 3$
41. $200,000 + 10,000 + 7000 + 500 + 80 + 6$ **43.** $50,000 + 900 + 40 + 3$ **45.** $200,000 + 500 + 80 + 3$
47. $400,000 + 3000 + 700 + 5$ **49.** 930 **51.** 1400 **53.** 7000 **55.** 44,000 **57.** 650,000

SECTION 1.2 pages 13–18

1. 11 **3.** 8 **5.** 19 **7.** 49 **9.** 28 **11.** 125 **13.** 693 **15.** 1099 **17.** 729 **19.** 1489
21. 14,399 **23.** 15,865 **25.** 156,499 **27.** 14,395 **29.** 1,786,688 **31.** 1,394,799 **33.** 18
35. 16,776 **37.** 149,999 **39.** 168,574 **41.** 129,843 **43.** 1584 **45.** 1219 **47.** 11,066
49. 5143 **51.** 102,317 **53.** 79,326 **55.** 1804 **57.** 1579 **59.** 19,740 **61.** 7420
63. 120,570 **65.** 207,453 **67.** 309,172 **69.** 343,263 **71.** 31 **73.** 534 **75.** 634
77. 14,302 **79.** 24,218 **81.** 11,974 **83.** 9323 **85.** 77,139 **87.** 10,417 **89.** 8217
91. Est.: 17,700 Cal.: 17,754 **93.** Est.: 2900 Cal.: 2872 **95.** Est.: 101,000 Cal.: 101,712
97. Est.: 158,000 Cal.: 158,763 **99.** Est.: 260,000 Cal.: 261,595 **101.** Est.: 940,000 Cal.: 946,718
103. Est.: 33,000,000 Cal.: 32,691,621 **105.** Est.: 34,000,000 Cal.: 34,420,922 **107.** The total attendance was 5792 people. **109.** The total number of yards gained by passing was 307. **111.** There were 8181 computers manufactured during the 3 months. **113a.** The total paid attendance at the sixth and seventh games was 131,348 people. **113b.** The total paid attendance for the entire series was 421,006 people.
115a. The total amount deposited was $1659. **115b.** The new checking account balance is $3007.
117. The total costs for the month were $5232.

SECTION 1.3 pages 23–26

1. 4 **3.** 4 **5.** 10 **7.** 7 **9.** 11 **11.** 9 **13.** 22 **15.** 60 **17.** 66 **19.** 33 **21.** 501
23. 962 **25.** 5002 **27.** 1513 **29.** 9 **31.** 7 **33.** 31 **35.** 47 **37.** 925 **39.** 4561
41. 3205 **43.** 1222 **45.** 1311 **47.** 3401 **49.** 1501 **51.** 65 **53.** 17 **55.** 8 **57.** 37
59. 353 **61.** 57 **63.** 160 **65.** 155 **67.** 24 **69.** 547 **71.** 45 **73.** 754 **75.** 378
77. 430 **79.** 3139 **81.** 3621 **83.** 738 **85.** 3545 **87.** 5258 **89.** 4887 **91.** 749
93. 5343 **95.** 66,463 **97.** 16,590 **99.** 52,404 **101.** 38,777 **103.** 4367 **105.** 4889
107. 2639 **109.** 1215 **111.** 5048 **113.** 628 **115.** 6532 **117.** 4286 **119.** 4042
121. 5209 **123.** 10,378 **125.** Est.: 60,000 Cal.: 60,427 **127.** Est.: 20,000 Cal.: 21,613
129. Est.: 360,000 Cal.: 358,979 **131.** Est.: 100,000 Cal.: 127,757 **133.** Est.: 500,000 Cal.: 456,767
135. Est.: 200,000 Cal.: 157,864 **137.** Est.: 1,000,000 Cal.: 1,082,767 **139.** Est.: 6,000,000 Cal.: 5,577,842 **141.** Est.: 2,000,000 Cal.: 1,625,369 **143.** There is $165 left in your checking account.
145. The amount that remains to be paid is $899. **147.** The odometer reading at the start of the trip was 1645 miles. **149a.** The total amount deducted was $276. **149b.** The steelworker's take home pay is $596.
151. There is $582 left in the budget. **153.** The team's profit for the season was $1,084,140.

SECTION 1.4 pages 31–34

1. 12 **3.** 35 **5.** 25 **7.** 0 **9.** 72 **11.** 21 **13.** 30 **15.** 54 **17.** 328 **19.** 220
21. 623 **23.** 711 **25.** 405 **27.** 280 **29.** 1143 **31.** 4010 **33.** 1500 **35.** 7248
37. 635 **39.** 2236 **41.** 5460 **43.** 1860 **45.** 1101 **47.** 1685 **49.** 46,963 **51.** 59,976
53. 19,120 **55.** 19,790 **57.** 144,759 **59.** 560,399 **61.** 651,000 **63.** 559,542 **65.** 336
67. 910 **69.** 640 **71.** 800 **73.** 1794 **75.** 1541 **77.** 63,063 **79.** 33,520 **81.** 380,834
83. 541,164 **85.** 400,995 **87.** 105,315 **89.** 66,000 **91.** 115,623 **93.** 428,770 **95.** 260,000
97. 344,463 **99.** 41,808 **101.** 189,500 **103.** 401,880 **105.** 1,052,763 **107.** 4,198,388
109. 1,232,000 **111.** 1,302,725 **113.** Est.: 540,000 Cal.: 550,935 **115.** Est.: 1,200,000 Cal.: 1,138,134 **117.** Est.: 600,000 Cal.: 605,022 **119.** Est.: 3,000,000 Cal.: 3,174,425
121. Est.: 18,000,000 Cal.: 19,212,228 **123.** Est.: 72,000,000 Cal.: 73,427,310 **125.** The plane used 4325

gallons of jet fuel. **127.** The machine can fill and cap 168,000 bottles of soft drink. **129.** The total cost of the car is $11,232. **131a.** The company can manufacture 192 video cameras. **131b.** There will be 4539 video cameras in stock. **133.** The total income was $824. **135.** There are 307,200 pixels on the screen. **137.** The oil company was purchased for $8,664,480,000.

SECTION 1.5 pages 43–46

1. 2   **3.** 6   **5.** 5   **7.** 16   **9.** 80   **11.** 210   **13.** 44   **15.** 206   **17.** 530   **19.** 902
21. 390   **23.** 1056   **25.** 2607   **27.** 9800   **29.** 21,560   **31.** 3580   **33.** 2 r1   **35.** 5 r2
37. 13 r1   **39.** 10 r3   **41.** 90 r2   **43.** 230 r1   **45.** 204 r3   **47.** 1347 r3   **49.** 778 r2   **51.** 391 r4
53. 2417 r1   **55.** 461 r8   **57.** 1160 r4   **59.** 708 r2   **61.** 3825   **63.** 9044 r2   **65.** 3 r15
67. 2 r3   **69.** 21 r36   **71.** 34 r2   **73.** 8 r8   **75.** 4 r49   **77.** 200 r25   **79.** 203 r2   **81.** 185 r2
83. 67 r70   **85.** 601 r8   **87.** 200 r9   **89.** 35 r47   **91.** 271   **93.** 4484 r6   **95.** 608   **97.** 15 r7
99. 1 r563   **101.** 50 r92   **103.** 40 r7   **105.** Est.: 5000 Cal.: 5129   **107.** Est.: 20,000 Cal.: 21,968
109. Est.: 22,500 Cal.: 24,596   **111.** Est.: 3000 Cal.: 2836   **113.** Est.: 3000 Cal.: 3024   **115.** Est.: 30,000 Cal.: 32,036   **117.** Est.: 80 Cal.: 83   **119.** Est.: 75 Cal.: 68   **121.** Each organization received $137,000.   **123.** Each disc can store 368,640 bytes of information.   **125.** The consultant charged $32 per hour.   **127a.** The remaining balance to be paid is $8976.   **127b.** The monthly payment is $187.
129. Each charity received $63,553.   **131.** There are 4216 employees participating in the plan.

SECTION 1.6 pages 49–50

1. 2^3 **3.** $6^3 \cdot 7^4$ **5.** $6^3 \cdot 11^5$ **7.** $3 \cdot 10^4$ **9.** $2^2 \cdot 3^3 \cdot 5^4$ **11.** $2 \cdot 3^2 \cdot 7^2 \cdot 11$ **13.** $2^2 \cdot 7^4 \cdot 11^4$
15. 8   **17.** 400   **19.** 900   **21.** 972   **23.** 120   **25.** 360   **27.** 0   **29.** 90,000   **31.** 540
33. 4050   **35.** 11,025   **37.** 25,920   **39.** 4,320,000   **41.** 92,160   **43.** 5   **45.** 10   **47.** 47
49. 8   **51.** 5   **53.** 8   **55.** 6   **57.** 53   **59.** 44   **61.** 19   **63.** 37   **65.** 168   **67.** 27
69. 14   **71.** 10   **73.** 9   **75.** 8   **77.** 7   **79.** 18

SECTION 1.7 pages 53–54

1. 1, 2, 4   **3.** 1, 2, 5, 10   **5.** 1, 5   **7.** 1, 2, 3, 4, 6, 12   **9.** 1, 2, 4, 8   **11.** 1, 13   **13.** 1, 2, 3, 6, 9, 18
15. 1, 5, 25   **17.** 1, 2, 3, 4, 6, 9, 12, 18, 36   **19.** 1, 2, 4, 7, 14, 28   **21.** 1, 29   **23.** 1, 2, 11, 22
25. 1, 2, 4, 11, 22, 44   **27.** 1, 7, 49   **29.** 1, 37   **31.** 1, 3, 19, 57   **33.** 1, 2, 3, 4, 6, 8, 12, 16, 24, 48
35. 1, 3, 29, 87   **37.** 1, 2, 23, 46   **39.** 1, 2, 5, 10, 25, 50   **41.** 1, 2, 3, 6, 11, 22, 33, 66
43. 1, 2, 4, 5, 8, 10, 16, 20, 40, 80   **45.** 1, 2, 3, 4, 6, 7, 12, 14, 21, 28, 42, 84   **47.** 1, 5, 17, 85
49. 1, 101   **51.** 1, 2, 3, 4, 5, 6, 8, 10, 12, 15, 20, 24, 30, 40, 60, 120   **53.** 1, 5, 25, 125
55. 1, 2, 3, 4, 6, 11, 12, 22, 33, 44, 66, 132   **57.** 1, 3, 5, 9, 15, 27, 45, 135   **59.** 1, 3, 7, 9, 21, 27, 63, 189
61. $2 \cdot 3$ **63.** prime   **65.** $2 \cdot 2 \cdot 2 \cdot 2$ **67.** $2 \cdot 2 \cdot 3$ **69.** $3 \cdot 3$ **71.** $2 \cdot 2 \cdot 3 \cdot 3$ **73.** prime
75. $2 \cdot 5 \cdot 5$ **77.** $5 \cdot 13$ **79.** $2 \cdot 2 \cdot 2 \cdot 2 \cdot 5$ **81.** $2 \cdot 3 \cdot 3$ **83.** $2 \cdot 2 \cdot 7$ **85.** prime   **87.** $2 \cdot 3 \cdot 7$
89. $3 \cdot 3 \cdot 3 \cdot 3$ **91.** $2 \cdot 11$ **93.** prime   **95.** $2 \cdot 5 \cdot 7$ **97.** $2 \cdot 43$ **99.** $5 \cdot 19$ **101.** prime
103. $5 \cdot 11$ **105.** $2 \cdot 61$ **107.** $2 \cdot 59$ **109.** $2 \cdot 3 \cdot 5 \cdot 5$ **111.** $2 \cdot 2 \cdot 2 \cdot 3 \cdot 5$ **113.** $2 \cdot 2 \cdot 2 \cdot 2 \cdot 5$
115. $7 \cdot 7 \cdot 7$ **117.** $2 \cdot 2 \cdot 2 \cdot 2 \cdot 5 \cdot 5$ **119.** $3 \cdot 3 \cdot 5 \cdot 5$

CHAPTER REVIEW pages 57–58

1. $101 > 87$ **2.** Two hundred seventy-six thousand fifty-seven   **3.** 2,011,044   **4.** $10,000 + 300 + 20 + 7$
5. 1081   **6.** 12,493   **7.** The total income from commissions was $2567.   **8.** The total amount deposited was $301. The new checking account balance is $817.   **9.** 1749   **10.** 5409   **11.** The odometer reading was 51,059 miles.   **12.** There is $396 left in your budget.   **13.** 22,761   **14.** 619,833   **15.** The total of the car payments is $1476.   **16.** The assistant's total pay for last week was $384.   **17.** 2135
18. 1306 r59   **19.** The car drove 27 miles per gallon of gasoline.   **20.** The monthly car payment is $155.
21. $2^4 \cdot 5^3$ **22.** $5^2 \cdot 7^5$ **23.** 600   **24.** 2   **25.** 17   **26.** 8   **27.** 1, 2, 3, 6, 9, 18   **28.** 1, 2, 3, 5, 6, 10, 15, 30   **29.** $2 \cdot 3 \cdot 7$ **30.** $2 \cdot 2 \cdot 2 \cdot 3 \cdot 3$

CHAPTER TEST pages 59–60

1. $21 > 19$ (1.1A)   **2.** Two hundred seven thousand sixty-eight (1.1B)   **3.** 1,204,006 (1.1B)
4. $900,000 + 6000 + 300 + 70 + 8$ (1.1C)   **5.** 75,000 (1.1D)   **6.** 96,798 (1.2A)   **7.** 135,915 (1.2B)
8a. They drove 855 miles.   **8b.** The odometer reading was 48,481 miles. (1.2C)   **9.** 9333 (1.3A)

10. 10,882 (1.3B)   **11.** The remaining balance is $13,955. (1.3C)   **12.** 726,104 (1.4A)
13. 6,854,144 (1.4B)   **14.** The investor will receive $2844 over 12 months. (1.4C)   **15.** 703 (1.5A)
16. 8710 r2 (1.5B)   **17.** 1121 r27 (1.5C)   **18.** The farmer needed 3000 boxes. (1.5D)
19. $3^3 \cdot 7^2$ (1.6A)   **20.** 432 (1.6A)   **21.** 9 (1.6B)   **22.** 7 (1.6B)   **23.** 1, 2, 4, 5, 10, 20 (1.7A)
24. $2 \cdot 2 \cdot 3 \cdot 7$ (1.7B)   **25.** $2 \cdot 3 \cdot 17$ (1.7B)

ANSWERS to Chapter 2 Odd-Numbered Exercises

SECTION 2.1   pages 65–66

1. 40   **3.** 24   **5.** 30   **7.** 12   **9.** 24   **11.** 60   **13.** 56   **15.** 9   **17.** 32   **19.** 36
21. 660   **23.** 9384   **25.** 24   **27.** 30   **29.** 24   **31.** 576   **33.** 1680   **35.** 72   **37.** 90
39. 300   **41.** 120   **43.** 1   **45.** 3   **47.** 5   **49.** 25   **51.** 1   **53.** 4   **55.** 4   **57.** 2
59. 48   **61.** 2   **63.** 8   **65.** 2   **67.** 1   **69.** 2   **71.** 3   **73.** 5   **75.** 8   **77.** 1   **79.** 25
81. 7   **83.** 8

SECTION 2.2   pages 69–70

1. $\frac{3}{4}$ **3.** $\frac{4}{7}$ **5.** $1\frac{1}{2}$ **7.** $2\frac{5}{8}$ **9.** $3\frac{3}{5}$ **11.** $\frac{5}{4}$ **13.** $\frac{8}{3}$ **15.** $\frac{28}{8}$ **17.** $2\frac{3}{4}$ **19.** 5   **21.** $1\frac{1}{8}$
23. $2\frac{3}{10}$ **25.** 3   **27.** $1\frac{1}{7}$ **29.** $2\frac{1}{3}$ **31.** 16   **33.** $2\frac{1}{8}$ **35.** $2\frac{2}{5}$ **37.** 1   **39.** 9   **41.** $\frac{7}{3}$
43. $\frac{13}{2}$ **45.** $\frac{41}{6}$ **47.** $\frac{37}{4}$ **49.** $\frac{21}{2}$ **51.** $\frac{73}{9}$ **53.** $\frac{58}{11}$ **55.** $\frac{21}{8}$ **57.** $\frac{13}{8}$ **59.** $\frac{100}{9}$ **61.** $\frac{27}{8}$
63. $\frac{85}{13}$

SECTION 2.3   pages 73–74

1. $\frac{5}{10}$ **3.** $\frac{9}{48}$ **5.** $\frac{12}{32}$ **7.** $\frac{9}{51}$ **9.** $\frac{12}{16}$ **11.** $\frac{27}{9}$ **13.** $\frac{20}{60}$ **15.** $\frac{44}{60}$ **17.** $\frac{12}{18}$ **19.** $\frac{35}{49}$ **21.** $\frac{10}{18}$
23. $\frac{21}{3}$ **25.** $\frac{35}{45}$ **27.** $\frac{60}{64}$ **29.** $\frac{21}{98}$ **31.** $\frac{30}{48}$ **33.** $\frac{15}{42}$ **35.** $\frac{102}{144}$ **37.** $\frac{153}{408}$ **39.** $\frac{340}{800}$ **41.** $\frac{30}{72}$
43. $\frac{119}{210}$ **45.** $\frac{1}{3}$ **47.** $\frac{1}{2}$ **49.** $\frac{1}{6}$ **51.** $\frac{2}{3}$ **53.** $1\frac{1}{2}$ **55.** 0   **57.** $\frac{9}{22}$ **59.** 3   **61.** $\frac{4}{21}$ **63.** $\frac{5}{11}$
65. $\frac{2}{9}$ **67.** $\frac{3}{4}$ **69.** $1\frac{1}{3}$ **71.** $\frac{3}{5}$ **73.** $\frac{1}{11}$ **75.** 4   **77.** $\frac{1}{3}$ **79.** $\frac{3}{5}$ **81.** $2\frac{1}{4}$ **83.** $\frac{1}{5}$ **85.** $\frac{19}{51}$
87. $\frac{27}{56}$

SECTION 2.4   pages 79–82

1. $\frac{3}{7}$ **3.** 1   **5.** $1\frac{4}{11}$ **7.** $3\frac{2}{5}$ **9.** $2\frac{4}{5}$ **11.** $2\frac{1}{4}$ **13.** $1\frac{3}{8}$ **15.** $1\frac{7}{15}$ **17.** $\frac{15}{16}$ **19.** $1\frac{4}{11}$
21. $1\frac{1}{6}$ **23.** $\frac{13}{14}$ **25.** $1\frac{1}{45}$ **27.** $\frac{17}{18}$ **29.** $\frac{35}{48}$ **31.** $\frac{23}{60}$ **33.** $\frac{56}{57}$ **35.** $\frac{13}{21}$ **37.** $\frac{33}{98}$ **39.** $\frac{37}{56}$
41. $1\frac{17}{18}$ **43.** $1\frac{11}{48}$ **45.** $1\frac{9}{20}$ **47.** $1\frac{109}{180}$ **49.** $1\frac{91}{144}$ **51.** $2\frac{5}{72}$ **53.** $1\frac{20}{33}$ **55.** $2\frac{11}{24}$ **57.** $5\frac{7}{10}$
59. $5\frac{11}{16}$ **61.** $8\frac{23}{60}$ **63.** $18\frac{8}{9}$ **65.** $11\frac{1}{36}$ **67.** $24\frac{11}{12}$ **69.** $6\frac{33}{40}$ **71.** $11\frac{19}{24}$ **73.** $12\frac{17}{22}$ **75.** $15\frac{29}{160}$
77. $20\frac{37}{48}$ **79.** $44\frac{17}{84}$ **81.** $11\frac{7}{18}$ **83.** $8\frac{1}{12}$ **85.** $12\frac{71}{105}$ **87.** $15\frac{1}{4}$ **89.** $12\frac{7}{24}$ **91.** $29\frac{227}{240}$
93. $14\frac{19}{24}$ **95.** The fractional amount of their income spent is $\frac{17}{24}$.   **97.** The price per share of the stock at the end of six months is $\$40\frac{3}{8}$.   **99.** The length of the shaft is $1\frac{5}{16}$ inches.   **101a.** The total hours worked is 20 hours.   **101b.** The total wages for the five days is $140.   **103.** The total snowfall for the year is $30\frac{7}{8}$ in.

SECTION 2.5 pages 87–90

1. $\frac{2}{17}$ 3. $\frac{1}{3}$ 5. $\frac{1}{10}$ 7. $\frac{7}{11}$ 9. $\frac{1}{4}$ 11. $\frac{2}{7}$ 13. $\frac{1}{2}$ 15. $\frac{19}{56}$ 17. $\frac{1}{2}$ 19. $\frac{4}{45}$ 21. $\frac{11}{18}$
23. $\frac{5}{48}$ 25. $\frac{1}{12}$ 27. $\frac{20}{57}$ 29. $\frac{17}{36}$ 31. $\frac{37}{51}$ 33. $\frac{17}{70}$ 35. $\frac{49}{120}$ 37. $\frac{25}{126}$ 39. $\frac{11}{48}$ 41. $\frac{23}{72}$
43. $\frac{81}{232}$ 45. $3\frac{1}{6}$ 47. $56\frac{4}{23}$ 49. $4\frac{1}{3}$ 51. $3\frac{2}{3}$ 53. $1\frac{3}{5}$ 55. $8\frac{2}{3}$ 57. $7\frac{43}{45}$ 59. $1\frac{2}{5}$ 61. $14\frac{4}{5}$
63. $20\frac{17}{24}$ 65. $3\frac{1}{8}$ 67. $25\frac{8}{15}$ 69. $6\frac{32}{45}$ 71. $2\frac{1}{160}$ 73. $1\frac{169}{200}$ 75. $48\frac{31}{70}$ 77. $85\frac{2}{9}$ 79. $9\frac{5}{13}$
81. $10\frac{169}{180}$ 83. $32\frac{43}{63}$ 85. $68\frac{1}{60}$ 87. $6\frac{83}{100}$ 89. The remaining piece of moulding is $5\frac{1}{3}$ feet long.
91. There are $2\frac{3}{4}$ hours remaining before landing. 93. The bridge shrank $\frac{1}{8}$ foot. 95. The missing dimension is $8\frac{19}{24}$ feet. 97a. The hikers will travel $17\frac{17}{24}$ miles the first two days. 97b. There are $9\frac{19}{24}$ miles left to hike. 99. The patient must lose $9\frac{1}{8}$ pounds.

SECTION 2.6 pages 95–98

1. $\frac{7}{12}$ 3. $\frac{7}{48}$ 5. $\frac{5}{12}$ 7. $\frac{1}{3}$ 9. $\frac{11}{20}$ 11. $\frac{1}{7}$ 13. $\frac{1}{8}$ 15. $\frac{5}{12}$ 17. $\frac{9}{50}$ 19. $\frac{5}{32}$ 21. 6
23. $\frac{2}{3}$ 25. $\frac{15}{44}$ 27. $\frac{2}{45}$ 29. $\frac{5}{14}$ 31. 10 33. $\frac{1}{15}$ 35. $\frac{10}{51}$ 37. $\frac{45}{328}$ 39. $\frac{10}{27}$ 41. 4
43. $\frac{100}{357}$ 45. $1\frac{1}{2}$ 47. 4 49. $\frac{4}{9}$ 51. $\frac{1}{2}$ 53. $16\frac{1}{2}$ 55. 10 57. $6\frac{3}{7}$ 59. $18\frac{1}{3}$ 61. $1\frac{5}{7}$
63. $3\frac{1}{2}$ 65. $25\frac{5}{8}$ 67. $1\frac{11}{16}$ 69. 16 71. 3 73. $8\frac{1}{4}$ 75. $27\frac{1}{2}$ 77. 7 79. $24\frac{1}{2}$ 81. 22
83. 33 85. $8\frac{37}{40}$ 87. $6\frac{19}{28}$ 89. $75\frac{5}{8}$ 91. $8\frac{1}{4}$ 93. $3\frac{1}{3}$ 95. 8 97. $8\frac{2}{3}$ 99. $19\frac{1}{2}$ 101. $3\frac{1}{2}$
103. 2 105. $7\frac{2}{5}$ 107. 16 109. $32\frac{4}{5}$ 111. $28\frac{1}{2}$ 113. 18 115. $8\frac{8}{15}$ 117. $34\frac{2}{5}$ 119. 519
121. 270 123. $80\frac{1}{3}$ 125. The car can travel $11\frac{3}{4}$ miles. 127. The car can travel 361 miles.
129. The cook should use 7 cups of flour. 131. The cost of 40 shares is $2735. 133a. The amount budgeted for housing and utilities is $1280. 133b. The amount remaining for other purposes is $1920.
135. The total cost of 24 choir robes is $504.

SECTION 2.7 pages 103–106

1. $\frac{5}{6}$ 3. $1\frac{5}{88}$ 5. 0 7. $1\frac{1}{3}$ 9. $8\frac{4}{5}$ 11. $2\frac{73}{256}$ 13. $1\frac{1}{9}$ 15. $2\frac{1}{13}$ 17. $\frac{2}{3}$ 19. $\frac{9}{10}$ 21. 2
23. 2 25. $\frac{1}{6}$ 27. 6 29. $\frac{1}{15}$ 31. 2 33. $2\frac{1}{2}$ 35. 3 37. 6 39. $\frac{1}{2}$ 41. 6 43. $20\frac{2}{3}$
45. $\frac{1}{30}$ 47. $1\frac{4}{5}$ 49. $8\frac{8}{9}$ 51. $\frac{1}{6}$ 53. $1\frac{1}{11}$ 55. $\frac{16}{21}$ 57. $\frac{7}{11}$ 59. $2\frac{13}{16}$ 61. $\frac{1}{9}$ 63. 8
65. $\frac{11}{28}$ 67. $\frac{5}{86}$ 69. 72 71. $\frac{3}{22}$ 73. $\frac{7}{44}$ 75. $\frac{33}{40}$ 77. $4\frac{4}{9}$ 79. $\frac{35}{47}$ 81. $53\frac{1}{2}$ 83. $10\frac{2}{7}$
85. $9\frac{39}{40}$ 87. $\frac{12}{53}$ 89. $2\frac{6}{43}$ 91. 0 93. $1\frac{31}{38}$ 95. $1\frac{4}{15}$ 97. $\frac{1}{144}$ 99. $\frac{10}{11}$ 101. 39
103. $2\frac{8}{39}$ 105. $13\frac{25}{49}$ 107. $4\frac{28}{61}$ 109. $4\frac{62}{191}$ 111. 68 113. 0 115. 6 117. $3\frac{13}{49}$
119. 4 121. $1\frac{61}{62}$ 123. $\frac{1}{2}$ 125. 800 boxes can be filled with cereal. 127. The sales executive can travel 25 miles on 1 gallon of gas. 129. The cost of the diamond would be $1920. 131. You can buy 60 shares of stock. 133a. There are $8\frac{1}{4}$ acres available for building. 133b. There will be 33 parcels of land available for sale. 135. 33 packages can be made from this shipment.

SECTION 2.8 pages 109–110

1. $\frac{11}{40} < \frac{19}{40}$ 3. $\frac{2}{3} < \frac{5}{7}$ 5. $\frac{5}{8} > \frac{7}{12}$ 7. $\frac{7}{9} < \frac{11}{12}$ 9. $\frac{13}{14} > \frac{19}{21}$ 11. $\frac{7}{24} < \frac{11}{30}$ 13. $\frac{9}{64}$ 15. $\frac{8}{729}$ 17. $\frac{2}{9}$
19. $\frac{3}{125}$ 21. $\frac{8}{245}$ 23. $\frac{1}{121}$ 25. $\frac{16}{1225}$ 27. $\frac{4}{49}$ 29. $\frac{9}{125}$ 31. $\frac{27}{88}$ 33. $\frac{4}{121}$ 35. $\frac{1}{30}$ 37. $1\frac{1}{2}$
39. $\frac{20}{21}$ 41. $\frac{12}{125}$ 43. $\frac{11}{32}$ 45. $\frac{17}{24}$ 47. $\frac{14}{15}$ 49. $2\frac{7}{10}$ 51. $\frac{21}{44}$ 53. $\frac{25}{39}$ 55. $\frac{5}{24}$ 57. $\frac{8}{9}$

CHAPTER REVIEW pages 113–114

1. 36 2. 54 3. 4 4. 5 5. $\frac{13}{4}$ 6. $1\frac{7}{8}$ 7. $3\frac{2}{5}$ 8. $\frac{19}{7}$ 9. 24 10. 32 11. $\frac{2}{3}$
12. $\frac{4}{11}$ 13. $1\frac{1}{8}$ 14. $1\frac{13}{18}$ 15. $5\frac{7}{8}$ 16. $18\frac{13}{54}$ 17. The total rainfall is $21\frac{7}{24}$ inches. 18. $\frac{1}{3}$
19. $\frac{25}{48}$ 20. $14\frac{19}{42}$ 21. $10\frac{1}{8}$ 22. The second checkpoint is $4\frac{3}{4}$ miles from the finish line. 23. $\frac{1}{15}$
24. $\frac{1}{8}$ 25. $9\frac{1}{24}$ 26. $16\frac{1}{2}$ 27. The car can travel 243 miles. 28. 2 29. $\frac{3}{4}$ 30. 2 31. $3\frac{1}{3}$
32. The cost of each acre was $36,000. 33. $\frac{11}{18} < \frac{17}{24}$ 34. $\frac{5}{16}$ 35. $\frac{5}{36}$ 36. $\frac{1}{15}$

CHAPTER TEST pages 115–116

1. 120 (2.1A) 2. 8 (2.1B) 3. $\frac{11}{4}$ (2.2A) 4. $3\frac{3}{5}$ (2.2B) 5. $\frac{49}{5}$ (2.2B) 6. $\frac{45}{72}$ (2.3A)
7. $\frac{5}{8}$ (2.3B) 8. $1\frac{11}{12}$ (2.4A) 9. $1\frac{61}{90}$ (2.4B) 10. $22\frac{4}{15}$ (2.4C) 11. The total rainfall for the 3 months was $21\frac{11}{24}$ inches. (2.4D) 12. $\frac{1}{4}$ (2.5A) 13. $\frac{7}{48}$ (2.5B) 14. $13\frac{81}{88}$ (2.5C) 15. The value of one share is $27\frac{7}{8}$. (2.5D) 16. $\frac{4}{9}$ (2.6A) 17. 8 (2.6B) 18. The electrician's total earnings are $420. (2.6C) 19. $1\frac{3}{7}$ (2.7A) 20. $2\frac{2}{19}$ (2.7B) 21. There were 11 lots available for sale. (2.7C)
22. $\frac{3}{8} < \frac{5}{12}$ (2.8A) 23. $\frac{1}{6}$ (2.8B) 24. $\frac{7}{24}$ (2.8C) 25. $\frac{5}{6}$ (2.8C)

CUMULATIVE REVIEW pages 117–118

1. 290,000 (1.1D) 2. 291,278 (1.3B) 3. 73,154 (1.4B) 4. 540 r12 (1.5C) 5. 1 (1.6B)
6. $2 \cdot 2 \cdot 11$ (1.7B) 7. 210 (2.1A) 8. 20 (2.1B) 9. $\frac{23}{3}$ (2.2B) 10. $6\frac{1}{4}$ (2.2B) 11. $\frac{15}{48}$ (2.3A)
12. $\frac{2}{5}$ (2.3B) 13. $1\frac{7}{48}$ (2.4B) 14. $14\frac{11}{48}$ (2.4C) 15. $\frac{13}{24}$ (2.5B) 16. $1\frac{7}{9}$ (2.5C) 17. $\frac{7}{20}$ (2.6A)
18. $7\frac{1}{2}$ (2.6B) 19. $1\frac{1}{20}$ (2.7A) 20. $2\frac{5}{8}$ (2.7B) 21. $\frac{1}{9}$ (2.8B) 22. $5\frac{5}{24}$ (2.8C) 23. There is $862 in the checking account at the end of the week. (1.3C) 24. The total income is $705. (1.4C)
25. The total weight is $12\frac{1}{24}$ pounds. (2.4D) 26. The length of the remaining piece is $4\frac{17}{24}$ feet. (2.5D)
27. The car can travel 225 miles. (2.6C) 28. 25 parcels can be sold from the remaining land. (2.7C)

ANSWERS to Chapter 3 Odd-Numbered Exercises

SECTION 3.1 pages 123–124

1. Twenty-seven hundredths 3. One and five thousandths 5. Thirty-six and four tenths
7. Thirty-five hundred-thousandths 9. Thirty-two and one hundred seven thousandths 11. Twelve and one thousand two hundred ninety-four ten-thousandths 13. Ten and seven thousandths 15. Fifty-two and ninety-five hundred-thousandths 17. Two hundred ninety-three ten-thousandths 19. Six and three hundred twenty-four thousandths 21. Two hundred seventy-six and three thousand two hundred ninety-seven ten-thousandths 23. Two hundred sixteen and seven hundred twenty-nine ten-thousandths

25. Four thousand six hundred twenty-five and three hundred seventy-nine ten-thousandths **27.** One and one hundred-thousandth **29.** 0.762 **31.** 0.000062 **33.** 8.0304 **35.** 304.07 **37.** 362.048 **39.** 3048.2002 **41.** 28.04375 **43.** 1308.0045 **45.** 7.4 **47.** 23.0 **49.** 22.68 **51.** 480.33 **53.** 1.039 **55.** 1946.375 **57.** 0.0299 **59.** 2 **61.** 0.010290

SECTION 3.2 pages 127–128

1. 150.1065 **3.** 95.8446 **5.** 69.644 **7.** 92.883 **9.** 113.205 **11.** 317.69398 **13.** 0.37 **15.** 3.107 **17.** 40.9820 **19.** 764.667 **21.** Est.: 608 Cal.: 608.245 **23.** Est.: 954 Cal.: 953.473 **25.** Est.: 1.5 Cal.: 1.47 **27.** Est.: 2.7 Cal.: 2.5906 **29.** The sales executive's monthly income is $2709.50. **31.** The length of the shaft is 5.65 inches. **33.** The total rainfall was 30.41 inches. **35a.** The car was driven 202.3 miles. **35b.** The odometer reading was 2019.7 miles. **37.** The total income was $310.77. **39.** The average number of points scored per game was 108 points.

SECTION 3.3 pages 131–132

1. 5.627 **3.** 113.6427 **5.** 6.7098 **7.** 215.697 **9.** 53.8776 **11.** 72.7091 **13.** 4.685 **15.** 10.0365 **17.** 0.014 **19.** 4.088 **21.** 0.7727 **23.** 3.273 **25.** 791.247 **27.** 547.951 **29.** 403.8557 **31.** 22.479 **33.** Est.: 600 Cal.: 590.25 **35.** Est.: 30 Cal.: 35.194 **37.** Est.: 3 Cal.: 2.74506 **39.** Est.: 7 Cal.: 7.14925 **41.** The amount of sales was $469.61. **43.** The assistant's raise was $0.49 per hour. **45.** The missing dimension is 2.59 feet. **47a.** The total amount of the checks written was $607.36. **47b.** The new checking account balance is $422.38. **49.** Rainfall was 0.86 inches below normal. **51.** The after-tax income was $3,699,644.14.

SECTION 3.4 pages 137–140

1. 0.36 **3.** 0.30 **5.** 0.25 **7.** 0.45 **9.** 6.93 **11.** 1.84 **13.** 4.32 **15.** 0.74 **17.** 39.5 **19.** 2.72 **21.** 0.603 **23.** 0.096 **25.** 13.50 **27.** 79.80 **29.** 4.316 **31.** 1.794 **33.** 0.06 **35.** 0.072 **37.** 0.1323 **39.** 0.03568 **41.** 0.0784 **43.** 0.076 **45.** 34.48 **47.** 580.5 **49.** 20.148 **51.** 0.04255 **53.** 0.17686 **55.** 0.19803 **57.** 14.8657 **59.** 0.0006608 **61.** 53.9961 **63.** 0.536335 **65.** 0.429 **67.** 2.116 **69.** 0.476 **71.** 1.022 **73.** 37.96 **75.** 2.318 **77.** 3.2 **79.** 6.5 **81.** 6285.6 **83.** 3200 **85.** 35,700 **87.** 6.3 **89.** 3.9 **91.** 49,000 **93.** 6.7 **95.** Est.: 90 Cal.: 91.2 **97.** Est.: 0.8 Cal.: 1.0472 **99.** Est.: 4.5 Cal.: 3.897 **101.** Est.: 12 Cal.: 11.2406 **103.** Est.: 0.32 Cal.: 0.371096 **105.** Est.: 30 Cal.: 31.8528 **107.** Est.: 2000 Cal.: 1941.0694 **109.** Est.: 0.00005 Cal.: 0.0000435 **111.** The cost to rent the car is $20.52. **113.** The amount received from recycling is $14.06. **115.** The broker's fee is $173.25. **117a.** The amount of the payments is $4590. **117b.** The total cost of the car is $6590. **119.** The total cost to rent the car is $64.20. **121.** The rocket will travel 639,360 miles in one day.

SECTION 3.5 pages 145–148

1. 0.82 **3.** 4.8 **5.** 89 **7.** 60 **9.** 0.06 **11.** 69.7 **13.** 84.3 **15.** 32.3 **17.** 5.06 **19.** 1.3 **21.** 0.11 **23.** 3.8 **25.** 10.5 **27.** 6.1 **29.** 680 **31.** 290 **33.** 3.6 **35.** 4.7 **37.** 1.3 **39.** 6.5 **41.** 0.8 **43.** 0.30 **45.** 0.08 **47.** 22.70 **49.** 0.55 **51.** 0.07 **53.** 0.249 **55.** 0.970 **57.** 0.101 **59.** 0.007 **61.** 4 **63.** 6 **65.** 4 **67.** 41 **69.** 116 **71.** 0.0039 **73.** 3.897 **75.** 2.37835 **77.** 82.547 **79.** 0.835 **81.** 8.765 **83.** 0.02954 **85.** 0.28932 **87.** 0.000139 **89.** Est.: 10 Cal.: 9.6944 **91.** Est.: 750 Cal.: 773.4940 **93.** Est.: 2 Cal.: 2.0069 **95.** Est.: 100 Cal.: 103.5556 **97.** Est.: 100 Cal.: 103.3774 **99.** Est.: 80 Cal.: 67.2037 **101.** There were 6.23 yards gained per carry. **103.** There were 102 gallons used. **105.** There can be 3 complete shelves cut from the board. **107a.** The amount to be paid in monthly payments is $842.58. **107b.** The amount of each monthly payment is $46.81. **109.** The store's total income from electric shavers is $7126.20. **111.** Each person would have to pay $3333.

SECTION 3.6 pages 151–152

1. 0.625 **3.** 0.667 **5.** 0.167 **7.** 0.417 **9.** 1.750 **11.** 1.500 **13.** 4.000 **15.** 0.003 **17.** 7.080 **19.** 37.500 **21.** 0.375 **23.** 0.208 **25.** 3.333 **27.** 5.444 **29.** 0.313 **31.** 2.214 **33.** $\frac{4}{5}$ **35.** $\frac{8}{25}$ **37.** $\frac{1}{8}$ **39.** $1\frac{1}{4}$ **41.** $16\frac{9}{10}$ **43.** $8\frac{2}{5}$ **45.** $8\frac{437}{1000}$ **47.** $2\frac{1}{4}$ **49.** $\frac{23}{150}$ **51.** $\frac{703}{800}$

53. $1\frac{17}{25}$ **55.** $\frac{9}{200}$ **57.** $16\frac{18}{25}$ **59.** $\frac{33}{100}$ **61.** $\frac{1}{3}$ **63.** $\frac{48}{175}$ **65.** $0.15 < 0.5$ **67.** $6.65 > 6.56$
69. $2.504 > 2.054$ **71.** $\frac{3}{8} > 0.365$ **73.** $\frac{2}{3} > 0.65$ **75.** $\frac{5}{9} > 0.55$ **77.** $0.62 > \frac{7}{15}$ **79.** $0.161 > \frac{1}{7}$
81. $0.86 > 0.855$ **83.** $1.005 > 0.5$ **85.** $\frac{5}{24} > 0.202$ **87.** $0.429 < \frac{4}{9}$

CHAPTER REVIEW pages 155–156

1. Twenty-two and ninety-two ten thousandths **2.** Three hundred forty-two and thirty-seven hundredths
3. 22.0092 **4.** 3.06753 **5.** 7.94 **6.** 0.05678 **7.** 36.714 **8.** 834.474 **9.** $478.02
10. 22.8635 **11.** 4.8785 **12.** The new checking account balance is $661.51. **13.** 8.932 **14.** 25.7446
15. The amount of income tax paid was $5600. **16.** 54.5 **17.** 6.594 **18.** The amount of each monthly payment is $123.45. **19.** 0.778 **20.** 2.33 **21.** $\frac{3}{8}$ **22.** $\frac{2}{3}$ **23.** $0.055 < 0.1$ **24.** $\frac{5}{8} > 0.62$

CHAPTER TEST pages 157–158

1. Forty-five and three hundred two ten-thousandths (3.1A) **2.** 209.07086 (3.1A) **3.** 7.095 (3.1B)
4. 0.0740 (3.1B) **5.** 255.957 (3.2A) **6.** 458.581 (3.2A) **7.** The total income was $1543.57. (3.2B)
8. 4.087 (3.3A) **9.** 27.76626 (3.3A) **10.** 1.37 inches (3.3B) **11.** 0.00548 (3.4A)
12. 2.4723 (3.4A) **13.** The cost of the long distance call is $4.63. (3.4B) **14.** 232 (3.5A)
15. 1.538 (3.5A) **16.** The amount of each monthly payment is $142.85. (3.5B) **17.** 0.692 (3.6A)
18. $\frac{33}{40}$ (3.6B) **19.** $0.66 < 0.666$ (3.6C) **20.** $\frac{3}{8} > 0.35$ (3.6C)

CUMULATIVE REVIEW pages 159–160

1. 235 r17 (1.5C) **2.** 128 (1.6A) **3.** 3 (1.6B) **4.** 72 (2.1A) **5.** $4\frac{2}{5}$ (2.2B) **6.** $\frac{37}{8}$ (2.2B)
7. $\frac{25}{60}$ (2.3A) **8.** $1\frac{17}{48}$ (2.4B) **9.** $8\frac{35}{36}$ (2.4C) **10.** $5\frac{23}{36}$ (2.5C) **11.** $\frac{1}{12}$ (2.6A) **12.** $9\frac{1}{8}$ (2.6B)
13. $1\frac{2}{9}$ (2.7A) **14.** $\frac{19}{20}$ (2.7B) **15.** $\frac{3}{16}$ (2.8B) **16.** $2\frac{5}{18}$ (2.8C) **17.** Sixty-five and three hundred nine ten-thousandths. (3.1A) **18.** 504.6991 (3.2A) **19.** 21.0764 (3.3A) **20.** 55.26066 (3.4A)
21. 2.154 (3.5A) **22.** 0.733 (3.6A) **23.** $\frac{1}{6}$ (3.6B) **24.** $\frac{8}{9} < 0.98$ (3.6C) **25.** There were 234 passengers on the continuing flight. (1.3C) **26.** The value of each share was $32\frac{7}{8}$. (2.5D) **27.** The checking account balance was $617.38. (3.3B) **28.** The resulting thickness of the bushing was 1.395 inches. (3.3B) **29.** The amount of income tax paid last year was $6008.80. (3.4B) **30.** The monthly payment is $23.87. (3.5B)

ANSWERS to Chapter 4 Odd-Numbered Exercises

SECTION 4.1 pages 165–166

1. $\frac{1}{5}$, 1:5, 1 TO 5 **3.** $\frac{2}{1}$, 2:1, 2 TO 1 **5.** $\frac{3}{8}$, 3:8, 3 TO 8 **7.** $\frac{37}{24}$, 37:24, 37 TO 24 **9.** $\frac{1}{1}$, 1:1, 1 TO 1
11. $\frac{7}{10}$, 7:10, 7 TO 10 **13.** $\frac{1}{2}$, 1:2, 1 TO 2 **15.** $\frac{2}{1}$, 2:1, 2 TO 1 **17.** $\frac{3}{4}$, 3:4, 3 TO 4 **19.** $\frac{5}{3}$, 5:3, 5 TO 3
21. $\frac{2}{3}$, 2:3, 2 TO 3 **23.** $\frac{2}{1}$, 2:1, 2 TO 1 **25.** $\frac{1}{1}$, 1:1, 1 TO 1 **27.** $\frac{1}{2}$, 1:2, 1 TO 2 **29.** $\frac{3}{2}$, 3:2, 3 TO 2
31. The ratio of housing cost to total income is $\frac{1}{3}$. **33.** The ratio of utilities cost to food cost is $\frac{3}{8}$. **35.** The ratio of people attending on Friday to people attending on Saturday was $\frac{16}{13}$. **37.** The ratio of turns in the primary coil to turns in the secondary coil is $\frac{1}{12}$. **39a.** The amount of increase is $20,000. **39b.** The ratio of the increase to the original value is $\frac{2}{9}$. **41.** The ratio of increase in price to original price is $\frac{2}{7}$.

SECTION 4.2 pages 169–170

1. $\frac{3 \text{ pounds}}{4 \text{ people}}$ 3. $\frac{\$20}{3 \text{ boards}}$ 5. $\frac{20 \text{ miles}}{1 \text{ gallon}}$ 7. $\frac{5 \text{ children}}{2 \text{ families}}$ 9. $\frac{8 \text{ gallons}}{1 \text{ hour}}$ 11. 2.5 feet/second   13. $325/week
15. 110 trees/acre   17. $4.71/hour   19. 52.4 miles/hour   21. 28 miles/gallon   23. $1.65/pound
25. $595.00/ton   27. The student drives 28.4 miles per gallon of gas.   29. The rocket uses 213,600 gallons of fuel per minute.   31. The dividend is $2.90 per share.   33a. There were 4878 compact discs meeting company standards.   33b. The cost was $5.44 per disc.   35. The profit was $0.60 per box of strawberries.
37. Light travels 186,000 miles per second.

SECTION 4.3 pages 175–178

1. True   3. Not true   5. Not true   7. True   9. True   11. Not true   13. True   15. True
17. True   19. Not true   21. True   23. Not true   25. Not true   27. True   29. $n = 3$
31. $n = 6$ 33. $n = 20$ 35. $n = 2$ 37. $n = 8$ 39. $n = 4$ 41. $n \approx 4.38$ 43. $n = 9.6$
45. $n \approx 10.67$ 47. $n \approx 6.67$ 49. $n = 26.25$ 51. $n = 96$ 53. $n \approx 9.78$ 55. $n \approx 3.43$
57. $n \approx 1.34$ 59. $n = 50.4$ 61. $n \approx 112.24$ 63. $n \approx 56.95$ 65. There are 168 calories in the serving of cereal.   67. There are 50 pounds of fertilizer used.   69. The monthly payment for the loan is $162.60.
71. The property tax is $2500.   73. It would take 2496 bricks.   75. The distance between the two cities is 114 miles.   77. 1.5 ounces of medication are required.   79. There would be 160,000 people voting in the election.   81. There would be 960 braking defects.   83. You own 400 shares of stock.   85. Students used the computer 300 hours.   87. 1750 pounds of salt can be recovered.

CHAPTER REVIEW pages 181–182

1. $\frac{2}{7}$, 2:7, 2 TO 7   2. $\frac{2}{5}$, 2:5, 2 TO 5   3. $\frac{1}{1}$, 1:1, 1 TO 1   4. $\frac{2}{5}$, 2:5, 2 TO 5   5. The ratio of the high temperature to the low temperature is 2:1.   6. The ratio of increase to original value is $\frac{1}{2}$.   7. The ratio of decrease in price to original price is $\frac{2}{5}$.   8. The ratio of T.V. advertising to newspaper advertising is $\frac{5}{2}$.
9. $\frac{100 \text{ miles}}{3 \text{ hours}}$ 10. $\frac{\$15}{4 \text{ hours}}$ 11. 62.5 miles/hour   12. $7.50/hour   13. 27.2 miles/gallon
14. $1.75/pound   15. The cost is $44.75 per share.   16. You drove an average of 56.8 miles/hour.
17. The cost is $.68/pound.   18. The cost per radio is $37.50.   19. True   20. True   21. True
22. Not true   23. $n = 36$ 24. $n = 68$ 25. $n = 65.45$ 26. $n = 19.44$ 27. There are 22.5 pounds of fertilizer used.   28. The cost of the insurance is $193.50.   29. The property tax is $2400.
30. It would take 1344 blocks to build the wall.

CHAPTER TEST pages 183–184

1. $\frac{3}{2}$, 3:2, 3 TO 2 (4.1A)   2. $\frac{3}{5}$, 3:5, 3 TO 5 (4.1A)   3. $\frac{1}{3}$, 1:3, 1 TO 3 (4.1A)   4. $\frac{1}{6}$, 1:6, 1 TO 6 (4.1A)
5. The ratio of the city temperature to the desert temperature is $\frac{43}{56}$. (4.1B)   6. The ratio of the cost of radio advertising to the total cost of advertising is $\frac{8}{13}$. (4.1B)   7. $\frac{9 \text{ supports}}{4 \text{ feet}}$ (4.2A)   8. $\frac{\$27}{4 \text{ boards}}$ (4.2A)
9. $1836.40/month (4.2B)   10. 30.5 miles/gallon (4.2B)   11. The cost of the lumber is $1.73/foot. (4.2C)   12. The plane's speed is 538 miles/hour. (4.2C)   13. True (4.3A)
14. Not true (4.3A)   15. $n = 40.5$ (4.3B)   16. $n = 144$ (4.3B)   17. The student's body contains 132 pounds of water. (4.3C)   18. The person requires 0.875 ounce of medication. (4.3C)   19. The property tax is $2800. (4.3C)   20. The dividend would be $625. (4.3C)

CUMULATIVE REVIEW pages 185–186

1. 9158 (1.3B)   2. $2^4 \cdot 3^3$ (1.6A)   3. 3 (1.6B)   4. $2 \cdot 2 \cdot 2 \cdot 2 \cdot 2 \cdot 5$ (1.7B)   5. 36 (2.1A)
6. 14 (2.1B)   7. $\frac{5}{8}$ (2.3B)   8. $8\frac{3}{10}$ (2.4C)   9. $5\frac{11}{18}$ (2.5C)   10. $2\frac{5}{6}$ (2.6B)   11. $4\frac{2}{3}$ (2.7B)
12. $\frac{23}{30}$ (2.8C)   13. Four and seven hundred nine ten-thousandths (3.1A)   14. 2.10 (3.1B)

Answers to Odd-Numbered Exercises **A37**

15. 1.990 (3.5A) **16.** $\frac{1}{15}$ (3.6B) **17.** $\frac{1}{8}$ (4.1A) **18.** $\frac{29¢}{2\text{ bars}}$ (4.2A) **19.** 33.4 miles/gallon (4.2B)
20. $n = 4.25$ (4.3B) **21.** 57.2 miles/hour (4.2C) **22.** $n = 36$ (4.3B) **23.** The new balance is $744. (1.3C) **24.** The monthly payment is $370. (1.5D) **25.** There are 105 pages remaining to be read. (2.6C) **26.** The cost for each acre is $36,000. (2.7C) **27.** The amount of change is $19.62. (3.3B)
28. The player's batting average is 0.271. (3.5B) **29.** There will be 25 inches eroded. (4.3C)
30. The person needs 1.6 ounces of medication. (4.3C)

ANSWERS to Chapter 5 Odd-Numbered Exercises

SECTION 5.1   pages 191–192

1. $\frac{1}{4}$, 0.25 **3.** $1\frac{3}{10}$, 1.30 **5.** 1, 1.00 **7.** $\frac{73}{100}$, 0.73 **9.** $3\frac{83}{100}$, 3.83 **11.** $\frac{7}{10}$, 0.70 **13.** $\frac{22}{25}$, 0.88
15. $\frac{8}{25}$, 0.32 **17.** $\frac{2}{3}$ **19.** $\frac{5}{6}$ **21.** $\frac{1}{9}$ **23.** $\frac{5}{11}$ **25.** $\frac{3}{70}$ **27.** $\frac{1}{15}$ **29.** 0.065 **31.** 0.0055
33. 0.0825 **35.** 0.0675 **37.** 0.0045 **39.** 0.804 **41.** 16% **43.** 5% **45.** 1% **47.** 70%
49. 124% **51.** 0.4% **53.** 0.6% **55.** 310.6% **57.** 54% **59.** 33.3% **61.** 62.5% **63.** 16.7%
65. 17.5% **67.** 177.8% **69.** 30% **71.** $23\frac{1}{3}$% **73.** $57\frac{1}{7}$% **75.** $27\frac{3}{11}$% **77.** $237\frac{1}{2}$% **79.** $216\frac{2}{3}$%

SECTION 5.2   pages 195–196

1. $n = 8$ **3.** $n = 10.8$ **5.** $n = 0.075$ **7.** $n = 80$ **9.** $n = 51.895$ **11.** $n = 7.5$ **13.** $n = 13$
15. $n = 3.75$ **17.** $n = 20$ **19.** $n = 210$ **21.** 5% of 95 **23.** 84% of 32 **25.** 15% of 80
27. $n \approx 25{,}805.032$ **29.** The amount deducted for income tax is $403.20. **31.** The profit is $4000.
33. The rebate is $1680. **35a.** The sales tax is $570. **35b.** The total cost including sales tax is $10,070.
37. The total number of employees needed is 671. **39.** The real estate taxes are $50,092.

SECTION 5.3   pages 199–200

1. $n = 32$% **3.** $n = 16\frac{2}{3}$% **5.** $n = 200$% **7.** $n = 37.5$% **9.** $n = 18$% **11.** $n = 0.25$%
13. $n = 20$% **15.** $n = 400$% **17.** $n = 2.5$% **19.** $n = 37.5$% **21.** $n = 0.25$% **23.** $n = 2.4$%
25. $n = 9.6$% **27.** $n \approx 43.01367$% **29.** The company spends 5% of its budget for advertising. **31.** The down payment is 20% of the selling price. **33.** The 1990 value is 180% of the 1984 value. **35a.** The price increase was $31,000. **35b.** The increase is 25% of the original price. **37.** Yes, the account executive answered enough questions to pass. **39.** The land area is 29% of the total area of the earth.

SECTION 5.4   pages 203–204

1. $n = 75$ **3.** $n = 50$ **5.** $n = 100$ **7.** $n = 85$ **9.** $n = 1200$ **11.** $n = 19.2$ **13.** $n = 7.5$
15. $n = 32$ **17.** $n = 200$ **19.** $n = 80$ **21.** $n = 9$ **23.** $n = 504$ **25.** $n = 108$ **27.** $n \approx 7122.1547$
29. The estimated life of the brakes is 50,000 miles. **31.** The city's population was 56,000 people.
33. The price at the competitor's store is $46.50. **35a.** The manufacturer tested 3000 computer boards.
35b. There were 2976 computer boards that were not defective. **37.** The increase in the number of boxes was 1680. **39.** The increase in earnings was $1,085,400.

SECTION 5.5   pages 207–208

1. $n = 65$ **3.** 25% **5.** $n = 75$ **7.** 12.5% **9.** $n = 400$ **11.** $n = 19.5$ **13.** 14.8%
15. $n = 62.62$ **17.** $n = 5$ **19.** $n = 45$ **21.** $n = 15$ **23.** 300% **25.** $n = 3200$ **27.** 2.4%
29. $n \approx 12{,}323.661$ **31.** The total amount collected is $24,500. **33.** The cost of the calculator was $71.25.
35. 12% of the alarms were false alarms. **37a.** The labor cost was $95.70. **37b.** The total cost of parts was $78.30. **39.** The percent of volunteers showing improvement was 92.5%. **41.** The new contract offered $1,464,375.

CHAPTER REVIEW pages 211–212

1. $\frac{3}{25}$ 2. $\frac{1}{6}$ 3. 0.42   4. 0.076   5. 38%   6. 150%   7. $n = 60$ 8. $n = 5.4$ 9. $n = 19.36$
10. $n = 77.5$ 11. The company spent $4500 for T.V. advertising.   12. The total cost is $1041.25.
13. $n = 150\%$ 14. $n = 20\%$ 15. $n = 7.3\%$ 16. $n = 613.3\%$ 17. The dividend is 4.6% of the stock price.   18. The percent increase in value was 64%.   19. $n = 75$ 20. $n = 10.9$ 21. $n = 157.5$
22. $n = 504$ 23. The city's population was 70,000 people.   24. The previous year's batting average was 0.252.   25. $n = 198.4$ 26. $n = 160\%$ 27. The computer cost $3000.   28. The student answered 85% of the questions correctly.

CHAPTER TEST pages 213–214

1. 0.973 (5.1A)   2. $\frac{1}{6}$ (5.1A)   3. 30% (5.1B)   4. 163% (5.1B)   5. 150% (5.1B)
6. $66\frac{2}{3}\%$ (5.1B)   7. $n = 50.05$ (5.2A)   8. $n = 61.36$ (5.2A)   9. 76% of 13 (5.2A)   10. 212% of 12 (5.2A)   11. The company spends $4500 for advertising. (5.2B)   12. There were 1170 pounds of vegetables unspoiled. (5.2B)   13. 25% (5.3A)   14. 216.67% (5.3A)   15. The store's temporary employees were 16% of the total number of permanent employees. (5.3B)   16. The student answered 91.3% of the questions correctly. (5.3B)   17. $n = 80$ (5.4A)   18. $n \approx 28.3$ (5.4A)   19. There were 32,000 transistors tested. (5.4B)   20. The increase is 60% of the original price. (5.4B)   21. $n \approx 143.0$ (5.5A)
22. 1000% (5.5A)   23. The increase in the hourly wage is $1.02. (5.5B)   24. The population now is 220% of the population 10 years ago. (5.5B)   25. The value of the car is $6500. (5.5B)

CUMULATIVE REVIEW pages 215–216

1. 4 (1.6B)   2. 240 (2.1A)   3. $10\frac{11}{24}$ (2.4C)   4. $12\frac{41}{48}$ (2.5C)   5. $12\frac{4}{7}$ (2.6B)   6. $\frac{7}{24}$ (2.7B)
7. $\frac{1}{3}$ (2.8B)   8. $\frac{13}{36}$ (2.8C)   9. 3.08 (3.1B)   10. 1.1196 (3.3A)   11. 34.2813 (3.5A)
12. 3.625 (3.6A)   13. $1\frac{3}{4}$ (3.6B)   14. $\frac{3}{8} < 0.87$ (3.6C)   15. $n = 53.3$ (4.3B)   16. $9.60/hour (4.2B)
17. $\frac{11}{60}$ (5.1A)   18. $83\frac{1}{3}\%$ (5.1B)   19. 19.56 (5.2A)   20. $133\frac{1}{3}\%$ (5.3A)   21. 9.92 (5.4A)
22. 342.9% (5.5A)   23. The mechanic's take-home pay is $592. (2.6C)   24. The monthly payment is $92.25. (3.5B)   25. There were 420 gallons of gasoline used during the month. (3.5B)   26. The real estate tax is $3000. (4.3C)   27. The sales tax is 6% of the purchase price. (5.3B)   28. 45% of the people surveyed did not favor the candidate. (5.3B)   29. The 1990 value is 250% of the 1985 value. (5.5B)
30. 18% of the children tested had levels of lead that exceeded federal standards. (5.5B)

ANSWERS to Chapter 6 Odd-Numbered Exercises

SECTION 6.1 pages 221–222

1. The unit cost is $0.056 per ounce.   3. The unit cost is $0.149 per ounce.   5. The unit cost is $0.033 per screw.   7. The unit cost is $0.195 per foot.   9. The unit cost is $0.037 per ounce.   11. The unit cost is $2.995 per clamp.   13. The more economical purchase is 12 ounces for $1.79.   15. The more economical purchase is 12 ounces for $2.64.   17. The more economical purchase is 46 ounces for $0.69.   19. The more economical purchase is 2 quarts for $4.79.   21. The more economical purchase is 200 tablets for $9.98.
23. The more economical purchase is 16 ounces for $1.32.   25. The total cost is $8.37.   27. The total cost is $2.94.   29. The total cost is $3.51.   31. The total cost is $0.40.   33. The total cost is $2.01.
35. The change is $5.25.

SECTION 6.2 pages 229–232

1. The percent increase is 5.5%.   3. The percent increase is 8%.   5. The percent increase is $33\frac{1}{3}\%$.
7a. The amount of increase is $2925.   7b. The new salary is $35,425.   9a. The amount of increase is $0.24.   9b. The total dividend per share of stock is $1.84.   11. The home is 240 square feet larger.

13. The markup is $71.25. **15.** The markup is $12.60. **17.** The markup rate is 30%. **19a.** The markup is $77.76. **19b.** The selling price is $239.76. **21a.** The markup is $21.70. **21b.** The selling price is $83.70. **23.** The selling price is $227.20. **25.** The percent decrease is 40%. **27.** The percent decrease is 40% **29.** The car loses $3360 in value. **31a.** The amount of the decrease is 12 cameras. **31b.** The percent decrease is 60%. **33a.** The amount of decrease was $15.20. **33b.** The average monthly bill is $60.80. **35.** The percent decrease is 20%. **37.** The discount rate is $33\frac{1}{3}$%. **39.** The discount is $68. **41.** The discount rate is 30%. **43a.** The discount is $7.20. **43b.** The sale price is $37.80. **45a.** The discount is $40. **45b.** The discount rate is 25%. **47.** The sale price is $35.70.

SECTION 6.3 pages 235–236

1. The simple interest due is $9000. **3.** The simple interest due is $13,500. **5.** The interest owed is $16. **7a.** The interest due is $72,000. **7b.** The monthly payment is $6187.50. **9a.** The interest charged is $1080. **9b.** The monthly payment is $545. **11.** The monthly payment is $1332.50. **13.** The amount in the account after one year is $820.62. **15.** The value of the investment is $134,253. **17.** The value of the investment is $11,235.93. **19a.** The value of the investment will be $111,446.25. **19b.** The interest earned will be $36,446.25. **21a.** The value of the investment will be $52,008.25. **21b.** Yes, the restaurant will have enough money in its account. **23.** The amount of interest earned is $15,967.58.

SECTION 6.4 pages 241–242

1. The mortgage is $82,450. **3.** The down payment is $7500. **5.** The down payment is $62,500. **7.** The loan origination fee is $3750. **9a.** The down payment is $4750. **9b.** The mortgage is $90,250. **11.** The mortgage is $86,400. **13.** The monthly mortgage payment is $1363.05. **15.** No, the couple cannot afford to buy the home. **17.** The monthly property tax is $53.75. **19a.** The monthly mortgage payment is $1960.22. **19b.** The interest is $1018.50 per month. **21.** The total monthly payment (mortgage and property tax) is $531.73. **23.** The monthly mortgage payment is $676.15.

SECTION 6.5 pages 245–246

1. The receptionist does not have enough money for the down payment. **3.** The sales tax is $742.50. **5.** The license fee is $250. **7a.** The sales tax is $420. **7b.** The total cost of the sales tax and the license fee is $595. **9a.** The down payment is $550. **9b.** The amount financed is $1650. **11.** The amount financed is $28,000. **13.** The monthly truck payment is $361.84. **15.** The cost to operate the car is $5120. **17.** The cost per mile is $0.11. **19.** The interest is $74.75. **21a.** The amount financed is $30,600. **21b.** The monthly truck payment is $650.16. **23.** The monthly car payment is $503.37.

SECTION 6.6 pages 249–250

1. The clerk earns $220. **3.** The commission earned is $2955. **5.** The commission received was $84. **7.** The teacher receives $3244 per month. **9.** The commission earned is $416. **11.** The commission earned was $540. **13.** The installer receives $280. **15.** The consultant's hourly wage is $125 per hour. **17a.** The therapist's hourly overtime wage is $21.56. **17b.** The therapist earns $344.96. **19a.** The increase in pay is $0.93 per hour. **19b.** The clerk's hourly wage is $7.13. **21a.** The amount of sales over $100,000 is $250,000. **21b.** The commission earned is $35,000. **23.** The attendant's hourly wage is $7.36.

SECTION 6.7 pages 257–260

1. The new balance is $486.32. **3.** The balance in the account is $1222.47. **5.** The current balance is $825.27. **7.** The current balance is $3000.82. **9.** Yes, there is enough money to purchase a refrigerator. **11.** Yes, there is enough money to make both purchases. **13.** The bank statement and checkbook balance. **15.** The bank statement and checkbook balance.

CHAPTER REVIEW pages 263–264

1. 14.5¢/ounce   **2.** 60 ounces for $8.40   **3.** The percent increase was 30.4%.   **4.** The yearly percent increase was 200%.   **5.** The selling price is $2079.   **6.** The percent increase in earnings is 15%.
7. The markup is $72.   **8.** The sale price is $141.   **9.** The simple interest due is $8250.
10. The simple interest due is $1650.   **11.** The value of the investment is $61,483.25.   **12.** The value of the investment will be $55,731.   **13.** The down payment is $18,750.   **14.** The loan origination fee is $1875.
15. The total monthly payment is $578.53.   **16.** The monthly mortgage payment is $1232.11.
17. The total cost is $1158.75.   **18.** The cost for the 4 items is 12.1¢/mile.   **19.** The amount of the interest is $97.33.   **20.** The monthly payment is $343.59.   **21.** The commission received was $3240.
22. The nurse's total income was $655.20.   **23.** The current balance is $943.68.   **24.** The current balance is $8866.58.

CHAPTER TEST pages 265–266

1. The cost per foot is $6.92. (6.1A)   **2.** The more economical purchase is 5 pounds for $1.65. (6.1B)
3. The cost is $14.53. (6.1C)   **4.** The percent increase is 20%. (6.2A)   **5.** The percent increase was 150%. (6.2A)   **6.** The selling price is $301. (6.2B)   **7.** The selling price is $6.25. (6.2B)   **8.** The percent decrease was 57.6%. (6.2C)   **9.** The percent decrease was 20%. (6.2C)   **10.** The sale price is $209.30. (6.2D)   **11.** The discount rate is 40%. (6.2D)   **12.** The simple interest due is $2750. (6.3A)
13. The interest earned was $43,055.70. (6.3B)   **14.** The loan origination fee is $3350. (6.4A)   **15.** The monthly mortgage payment is $1108.61. (6.4B)   **16.** The amount financed is $14,350. (6.5A)   **17.** The monthly car payment is $320.02. (6.5B)   **18.** The nurse earns $703.50. (6.6A)   **19.** The current balance is $6612.25. (6.7A)   **20.** The bank statement and checkbook balance. (6.7B)

CUMULATIVE REVIEW pages 267–268

1. 13 (1.6B)   **2.** $8\frac{13}{24}$ (2.4C)   **3.** $2\frac{37}{48}$ (2.5C)   **4.** 9 (2.6B)   **5.** 2 (2.7B)   **6.** 5 (2.8C)
7. 52.2 (3.5A)   **8.** 1.417 (3.6A)   **9.** $51.25/hour (4.2B)   **10.** $n = 10.94$ (4.3B)   **11.** 62.5% (5.1B)
12. 27.3 (5.2A)   **13.** 0.182 (5.1A)   **14.** 42% (5.3A)   **15.** 250 (5.4A)   **16.** 154.76 (5.5A)
17. The total rainfall was $13\frac{11}{12}$ inches. (2.4D)   **18.** The amount the family paid in taxes was $570. (2.6C)
19. The ratio of decrease in price to original price is $\frac{3}{5}$. (4.1B)   **20.** The car drove 33.4 miles per gallon. (4.2C)   **21.** The unit cost is $0.93. (4.2C)   **22.** The dividend is $280. (4.3C)   **23.** The sale price is $720. (5.2B)   **24.** The selling price is $119. (6.2B)   **25.** The percent increase is 8%. (6.2A)
26. The simple interest due is $6000. (6.3A)   **27.** The monthly payment is $381.60. (6.5B)   **28.** The new balance is $2243.77. (6.7A)   **29.** The cost per mile is $0.20. (6.5B)   **30.** The monthly mortgage payment is $743.18. (6.4A)

Index

Addend, 9
Addition
  Associative Property of, 9
  carrying, 10-11
  Commutative Property of, 9
  of decimals, 125-128
  defined, 9
  of fractions, 75-82
  and large numbers, 10
  of mixed numbers, 76-77
  on the number line, 9
  and Order of Operations Agreement, 48
  Property of Zero, 9
  related to subtaction, 19
  table of basic facts, A2
  of whole numbers, 9-18
Application problems
  bank statements, 257-260
  car expenses, 245-246
  decimals, 126, 128, 130, 132, 136, 140, 144, 148
  division, 41-42, 46
  fractions, 78, 81-82, 86, 90, 94, 98, 101-102, 106
  interest, 235-236
  multiplication, 30, 34
  percent decrease, 231-232
  percent increase, 229-230
  percents, 194, 196, 198, 200, 201-202, 204, 206, 208
  proportions, 174, 177-178
  purchasing, 221-222
  rates, 168, 170
  ratios, 164, 166
  real estate expenses, 241-242
  strategy for, 12
  subtraction, 22, 26
  wages, 249-250
  whole numbers, 12, 17-18
Approximately equal to ($\approx$), 11
Associative Property
  of Addition, 9
  of Multiplication, 27

Balancing a checkbook, 252-256
Bank statements, 251-260
Base, in percent equations, 201-204
Basic percent equation, 193
Borrowing in subtraction
  mixed numbers, 84
  whole numbers, 20-21

Business and consumer applications, 217-268
  bank statements, 251-260
  car expenses, 243-246
  interest, 233-236
  percent decrease, 226-228
  percent increase, 223-225
  purchasing, 219-222
  real estate expenses, 237-242
  wages, 247-250

Calculators
  and estimation, 11
  and fractions, 153
  memory key, 179
  and Order of Operations Agreement, 55
  percent key, 209
  and rates, 179
  and unit rates, 179
Car
  owning, 244
  purchasing, 243
Carrying in addition, 10-11
Chapter reviews
  business and consumer applications, 263-264
  decimals, 155-156
  fractions, 113-114
  percents, 211-212
  ratio and proportion, 181-182
  whole numbers 57-58
Chapter summaries, 55-56, 111-112, 153-154, 180, 210, 261-262
Chapter tests, 59-60, 115-116, 157-158, 183-184, 213-214, 265-266
Checkbook, balancing, 251-256
Commission, 247
Common denominator, 75
Common factor, 64
Common multiple, 63
Commutative Property
  of Addition, 9
  of Multiplication, 27
Comparison
  rates, 167-170
  ratios, 163-166
  unit costs, 219-220
Composite number, 52
Compound interest, 234
  table of, A4-A5

Computers
  and amortization, 261
  and fractions, 111
Consumer applications, *see* Business and consumer applications
Conversion
  decimals to fractions, 149-150
  fractions to decimals, 149
  of percents, 189-190
Cost, 224
  total, 220
  unit, 219-220
Cross products, 171
Cube of a number, 47
Cumulative reviews, 117-118, 159-160, 185-186, 215-216, 267-268

Data, 271
Decimal fractions, 120
Decimal notation, 121
Decimal point, 121
Decimals, 119-160
  addition of, 125-128
  converting to/from fractions, 149-150
  division of, 141-148
  estimation of, 125, 130, 135, 143
  multiplication of, 133-140
  and number line, 150
  order relation, 150
  and percents, 189-190
  and power of 10, 133-135
  rounding, 122, 149
  standard form, 121
  subtraction of, 129-132
  word form, 121
  writing, 121-122
Decrease, percent, 226-228
Denominator, 67
  common, 75
  least common, 75, 83
Difference, 19
Digits, rounding, 6
Discount, 227-228
Discount rate, 227-228
Dividend, 35
Division
  of decimals, 141-148
  defined, 35
  estimation of, 41
  and factors, 51
  of fractions, 99-107

and large numbers, 39
of mixed numbers, 99-100
and Order of Operations
  Agreement, 48
by powers of ten, 133-134
Properties of Zero in, 35
as related to multiplication, 35
rounding, 141
of whole numbers, 35-46
writing as fractions, 133
Divisor, 35
Down payment
  car purchase, 243
  house purchase, 237

Egyptian fractions, 62
Equivalent fractions, 71
Estimation
  adding decimals, 125
  calculators, 11
  dividing decimals, 143
  division, 41
  multiplication, 29
  multiplying decimals, 135
  subtracting decimals, 130
  subtraction, 21
  whole numbers, 11
Expanded form of a number, 4-5
Exponential notation, 47
Exponents, 47
  and fractions, 107-108
  and Order of Operations
    Agreement, 48
  simplifying expressions
    containing, 107-108
Expressions, and the Order of
  Operations Agreement, 48

Factor(s), 51
  common, 64
  greatest common, 64
  in multiplication, 27
Factorization, prime, 52, 63, 64
Fraction bar, 67
Fractions, 61-118
  addition of, 75-82
  and calculators, 153
  and computers, 111
  converting to/from decimals,
    133, 149
  decimal, 120
  division of, 99-107
  Egyptian, 62
  equivalent, 71
  and exponents, 107-108
  and greatest common factor, 64
  improper, 67-68
  inverting, 99
  and least common multiple, 63
  mixed numbers, see Mixed
    numbers

multiplication of, 91-98
and Multiplication Property of
  One, 71
on the number line, 107
and Order of Operations
  Agreement, 108
order relation, 107
and percents, 189
proper, 67
reciprocal of, 99
simplest form, 72
subtraction of, 83-90
writing, 67
Frequency polygon, 284

Graph(s)
  on the number line, 3, 107
  whole number, 3
Greater than, 3
Greatest common factor (GCF), 64

Hindu-Arabic system, 2
Home
  owning, 238-240
  purchasing, 237-238
Hourly wage, 247

Improper fractions, 67-68
  and mixed numbers, 68
  and whole numbers, 68
Increase, percent, 223-225
Interest
  compound, 234
  simple, 233
Interest rate, 233
Inverting fractions, 99

Least common denominator
  (LCD), 75, 83
Least common multiple (LCM), 63
Less than, 3
Loan
  amortization of, 261
  origination fee, 237-238

Markup, 224-225
Markup rate, 224-225
Memory key, on calculators, 179
Minuend, 19
Mixed numbers, 67
  addition of, 76-77
  division of, 99-100
  and improper fractions, 68
  multiplication of, 92-93
  subtraction of, 84-85
Monthly payment table, A6
Mortgage, 237-240
Multiple
  common, 63
  least common, 63

Multiplication
  Associative Property of, 27
  Commutative Property of, 27
  of decimals, 133-140
  defined, 27
  estimation of, 29
  of fractions, 91-98
  large whole numbers, 28-29
  of mixed numbers, 92-93
  number line and, 27
  Order of Operations Agreement
    and, 48
  powers of ten, 47, 133
  Property of One, 27, 71, 72
  Property of Zero, 27
  table of basic facts, A2
  of whole numbers, 27-34
  zeros in, 28-29
Musical scales, 162

Number line, 3
  addition, 9
  decimals, 150
  fractions, 107
  graphs on, 3, 107
  multiplication, 27
  subtraction, 19
  whole numbers, 3
Number(s)
  composite, 52
  cubed, 47
  decimal notation, 121
  expanded form of, 4-5
  factoring, 51-54
  family tree for, 2
  fractions, see Fractions
  large, 10
  mixed, see Mixed numbers
  multiples of, 63
  prime, 52-54
  squared, 47
  whole, see Whole numbers
Numerator, 67

One, Multiplication Property of,
  27, 71, 72
Order of Operations Agreement, 48
  and calculators, 55
  fractions, 108
Order relation
  decimals, 150
  fractions, 107
  whole numbers, 3

Percent, 187-216
  commission, 247
  and decimals, 189-190
  of decrease, 226-228
  defined, 189
  discount, 227-228
  and fractions, 189

home loans, 237-238
of increase, 223-225
interest, 233
ratios, 205
symbol for, 188
writing, 189-190
Percent equations
basic, 193
proportion method, 205-208
solving for amount, 193-196
solving for base, 201-204
solving for percent, 197-200
Percent key, on calculator, 209
Period, 4
Place value
decimals, 121, 122, 125, 129
expressed as a power of ten, 47
rounding to a given, 5-6, 122
and subtraction, 20
whole numbers, 4
Place-value chart, 4
decimals, 125
division, 36
expanded form, 4
multiplication, 29
whole numbers, 4-5
Points, 237-238
Powers of ten, 47
and decimals, 133-135
and division, 133-134
and multiplication, 47, 133
and place value, 47
Prime factorization, 52
greatest common factors and, 64
least common multiples and, 63
Prime number, 52-54
Principal, 233
Product, 27
cross, 171
Proper fraction, 67
Properties
of addition, 9
Associative
of Addition, 9
of Multiplication, 27
Commutative
of Addition, 9
of Multiplication, 27
of One, Multiplication, 27, 71, 72
of Zero
in addition, 9
in division, 35
in multiplication, 27
Property tax, 239
Proportions, 171-178
cross products, 171
defined, 171

not true, 171
and percent problems, 205-208
solving, 172-173
true, 171
Purchasing, 219-222

Quantity, comparison of, 163-170
Quotient, 35, 37-38

Rate(s), 167-170
and calculators, 179
and proportion, 171
simplest form, 167
unit, 167, 179
writing, 167
Ratio(s), 163-166
percents, 205
pictographs and, 271
proportion and, 171
simplest form, 163
writing, 163
Real estate expenses, 237-242
Reciprocals, 99
Remainder, 37-38
Rounding
decimals, 122, 149
in division, 141
and place value, 5-6, 122
whole numbers, 5-6, 29

Salary, 247
Sale price, 227-228
Selling price, 224-225
Simple interest, 233
Simplest form rate, 167
Simplest form ratio, 163
Simplifying fractions, 72
Solving proportions, 172-173
Square of a number, 47
Standard form
decimal, 121
whole number, 4
Subtraction
borrowing, 20-21
checking, 19
of decimals, 129-132
defined, 19
estimation of, 21
of fractions, 83-90
of mixed numbers, 84-85
on the number line, 19
Order of Operations Agreement and, 48
related to addition, 19
of whole numbers, 19-26
zero and, 20
Subtrahend, 19
Sum, 9

Symbols
approximately equal to ($\approx$), 11
division ($\div$, $-$), 36
greater than ($>$), 3
less than ($<$), 3
multiplication ($\times$, $\cdot$), 27
percent (%), 188

Tables of basic facts
addition, A2
compound interest, A4-A5
monthly payment, A6
multiplication, A2
square roots, A3
T-diagram, 52
Ten, powers of, see Powers of ten
Total cost, 220

Unit, 163
Unit cost, 219-220
Unit rate, 167
and calculators, 179

Wage, 247-250
Whole numbers, 1-60
addition of, 9-18
borrowing, 20-21
carrying, 10-11
definition of, 3
division of, 35-46
estimating quotient of, 41
estimation of, 11
expanded form, 4-5
graph of, 3
and improper fractions, 68
multiplication of, 27-34
on the number line, 3
Order of Operations Agreement, 48
order relation, 3
reading, 4
reciprocal of, 99
rounding, 5-6, 29
standard form, 4
subtraction of, 19-26
word form, 4
writing, 4-5
Word form
decimals, 121
whole numbers, 4

Zero
Addition Property of, 9
Multiplication Property of, 27
properties of, in division, 35
and subtraction with borrowing, 20